Polyhydroxyalkanoate (PHA) Based Blends, Composites and Nanocomposites

RSC Green Chemistry

Editor-in-Chief:
Professor James Clark, *Department of Chemistry, University of York, UK*

Series Editors:
Professor George A Kraus, *Department of Chemistry, Iowa State University, Ames, Iowa, USA*
Professor dr. ir. Andrzej Stankiewicz, *Delft University of Technology, The Netherlands*
Professor Peter Siedl, *Federal University of Rio de Janeiro, Brazil*
Professor Yuan Kou, *Peking University, China*

How to obtain future titles on publication:
A standing order plan is available for this series. A standing order will bring
delivery of each new volume immediately on publication.

For further information please contact:
Book Sales Department, Royal Society of Chemistry, Thomas Graham House,
Science Park, Milton Road, Cambridge, CB4 0WF, UK
Telephone: +44 (0)1223 420066, Fax: +44 (0)1223 420247
Email: booksales@rsc.org
Visit our website at www.rsc.org/books

Polyhydroxyalkanoate (PHA) Based Blends, Composites and Nanocomposites

Edited by

Ipsita Roy
University of Westminster, London, UK
Email: I.Roy01@westminster.ac.uk

Visakh P M
Mahatma Gandhi University, Kottayam, India
Email: visagam143@gmail.com

THE QUEEN'S AWARDS
FOR ENTERPRISE:
INTERNATIONAL TRADE
2013

RSC Green Chemistry No. 30

Print ISBN: 978-1-84973-946-7
PDF eISBN: 978-1-78262-231-4
ISSN: 1757-7039

A catalogue record for this book is available from the British Library

Published by The Royal Society of Chemistry,
Thomas Graham House, Science Park, Milton Road,
Cambridge CB4 0WF, UK

Registered Charity Number 207890

For further information see our web site at www.rsc.org

Printed and bound by CPI Group (UK) Ltd, Croydon, CR0 4YY

Preface

This book on "Polyhydroxyalkanoates (PHAs)-based Blends, Composites and Nanocomposites" summarizes many of the recent research accomplishments in the area of polyhydroxyalkanoates such as state-of-the-art polyhydroxyalkanoates, their blends, composites and nanocomposites, the state of the art, new challenges, opportunities for polyhydroxyalkanoates and their structure, properties and sources, recovery of polyhydroxyalkanoates, blends of polyhydroxyalkanoates, nanocomposites of polyhydroxyalkanoates, polyhydroxyalkanoates-based multiphase materials, modification of polyhydroxyalkanoates, polyhydroxyalkanoates as packaging materials, current applications and future prospects of polyhydroxyalkanoates. As the title indicates, the book emphasizes the various aspects of polyhydroxyalkanoates and their blends, composites and nanocomposites. This book is intended to serve as a "one stop" reference resource for important research accomplishments in this area. This book will be a very valuable reference source for university and college faculties, professionals, post-doctoral research fellows, senior graduate students, researchers from R&D laboratories working in the area of polyhydroxyalkanoates and their blends, composites and nanocomposites. The various chapters in this book are contributed by prominent researchers from industry, academia and government/private research laboratories across the globe. It covers an up-to-date record of the major findings and observations in the field of polyhydroxyalkanoates and their blends, composites and nanocomposites.

The first chapter on polyhydroxyalkanoates and their blends, composites and nanocomposites covers the state of the art, new challenges, and opportunities for polyhydroxyalkanoates based studies and research. The second chapter provides an overview of polyhydroxyalkanoates: their structure, properties and sources. The third chapter concentrates mainly on the recovery of polyhydroxyalkanoates. The chapter's authors discuss the

RSC Green Chemistry No. 30
Polyhydroxyalkanoate (PHA) Based Blends, Composites and Nanocomposites
Edited by Ipsita Roy and Visakh P M
© The Royal Society of Chemistry 2015
Published by the Royal Society of Chemistry, www.rsc.org

different recovery methods, such as chemical methods, biological methods, mechanical methods, physical methods, combined methods and others. The authors also discuss the characterization of recovered polyhydroxyalkanoates. They also discuss the following sub-topics: solvent extraction, chemical digestion, enzymatic digestion, bead mill, ultrasonication, osmotic shockfreezing.

A survey on polyhydroxyalkanoate blends is presented in the fourth chapter; the authors describe the production, characterisation and applications of polyhydroxyalkanoate blends. Different types of blends, such as PHAs/PLA blends PHAs/PCL blends, PHAs/cellulose derivatives, PHAs/starch derivatives and PHAs/chitosan, are discussed. The fifth chapter discusses the nanocomposites of polyhydroxyalkanoates, including the preparation, characterisation and applications of different types of polyhydroxyalkanoates-based nanocomposites. The preparation methods discussed include solution intercalation, melt intercalation and *in situ* intercalative polymerization. The sixth chapter deals with polyhydroxyalkanoates-based multiphase materials. The chapter explains many properties such as renewability, sustainable development, biodegradability, compostability, classification, chemical structures and the crystalline structures of polyhydroxyalkanoates. The authors explain other interesting topics such as PHA-based multiphase materials, generalities, plasticization, nucleating agents, PHA-based blends, PHA-based multilayers, PHA-based biocomposites, PHA-based nano-biocomposites, production and some applications, the industrial production of PHA-based materials and PHA-based materials for biomedical applications.

The seventh chapter discusses the modification of polyhydroxyalkanoates. From this chapter, we can see different kinds of modification methods such as chemical modification, physical modification and modification with enzymes.

The section on the chemical modification of PHAs is divided into different topics such as carboxylation, hydroxylation, epoxidation, halogenations and graft copolymerization. The section on the physical modification of PHAs is divided into different topics such as PHA blending and coating, PHA irradiation, ion implantation, plasma treatment, electrospinning. To conclude, the authors discuss the modification of PHAs with enzymes.

Polyhydroxyalkanoates as packaging materials, current applications and future prospects are discussed in the eighth chapter. In this chapter, the authors discuss different topics such as the need for sustainable packaging materials, biodegradable plastics, factors affecting PHA production, properties of PHAs for packaging, PHAs for food and industrial packaging and PHAs for medical packaging.

The ninth chapter is on the packaging applications of polyhydroxyalkanoates (PHAs). The authors discuss many topics such as polyhydroxyalkanoates, fundamental bioprocess steps for obtaining PHAs, degradation PHA uses in food packaging, methods for obtaining plastic products and examples of applications of bioplastics in food packaging *etc.*

Finally, the editors would like to express their sincere gratitude to all the contributors to this book, who provided excellent support to the successful completion of this venture. We are grateful to them for the commitment and the sincerity they have shown towards their contribution in the book. Without their enthusiasm and support, the compilation of a book would not have been possible. We would like to thank all the reviewers who have given their valuable time to make critical comments on each chapter. We also thank the publisher—the Royal Society of Chemistry—for recognizing the demand for such a book, and for realizing the increasing importance of the area of "Polyhydroxyalkanoates (PHAs)-based Blends, Composites and Nanocomposites" and for starting this new project.

Ipsita Roy
Visakh P.M.

Contents

RSC Green Chemistry No. 30
Polyhydroxyalkanoate (PHA) Based Blends, Composites and Nanocomposites
Edited by Ipsita Roy and Visakh P M
© The Royal Society of Chemistry 2015
Published by the Royal Society of Chemistry, www.rsc.org

CHAPTER 1

Polyhydroxyalkanoates (PHAs), their Blends, Composites and Nanocomposites: State of the Art, New Challenges and Opportunities

VISAKH P. M.

Senior Researcher, School of Chemical Sciences, Mahatma Gandhi University, Kottayam, Kerala, India
Email: visagam143@gmail.com

1.1 Polyhydroxyalkanoates: Structure, Properties and Sources

Polyhydroxyalkanoates are biopolyesters with various side chains and fatty acids with hydroxyl groups at the 4- or 5-position. They consist of (R)-3-hydroxy fatty acids. There are three types of polyhydroxyalkanoates: (a) short chain length hydroxyalkanoic acids (PHA_{SCL}) with an alkyl side chain, which are produced by *Ralstonia eutropha* and many other bacteria. PHA_{SCL} contain 3–5 carbon atoms, for example poly-3-hydroxybutyrate (P3HB), poly-4-hydroxybutyrate (P4HB); (b) medium chain length hydroxyalkanoic acids (PHA_{MCL}) with alkyl side chains that are produced by *Pseudomonas oleovorans* and other *Pseudomonas sensu stricto*. PHA_{MCL} contain 6–14 carbon atoms and (c) long chain length (PHA_{LCL}) obtained from long chain fatty

RSC Green Chemistry No. 30
Polyhydroxyalkanoate (PHA) Based Blends, Composites and Nanocomposites
Edited by Ipsita Roy and Visakh P M
© The Royal Society of Chemistry 2015
Published by the Royal Society of Chemistry, www.rsc.org

acids, which contain more than 14 carbon atoms The monomer composition, macromolecular structure and physical chemical properties of PHAs vary, depending on the producer organism as well as on the carbon source used for the growth.[1-3] PHAs containing double bonds can also be produced by recombinant *Methylobacterium extorquens* strains when fed unsaturated fatty acids. These PHAs comprise PHA_{SCL} and PHA_{MCL}. One reason for choosing a methylotrophic microorganism for such a purpose was that an important portion of the production process would use methanol. In spite of PHB being considered an environmentally friendly polymer with similar material properties to polypropylene (PP), it has not been used on a large scale to replace conventional polymers because it presents some drawbacks in its mechanical properties. Considering polymer mechanical properties, it is important to consider three basic properties when comparing the usefulness of a polymer for a given commodity application. It is hard to process PHB due to its high melting temperature of approximately 170 °C, which is very near to its degradation temperature. Therefore, a solution to these drawbacks could be the copolymerization of 3-HB with other monomers that confer less stiffness and tougher properties (which bestow greater flexibility and lessen breakage) and to reduce the melting point.

The monomer composition of PHA has considerable effects on its physical properties.

The PHA structure can effectively be controlled by adjusting the carbon substrates to achieve desired monomer contents, by engineering metabolic pathways in the hosts or by feeding the culture with carbon substrates containing functional side chains that in a second step can suffer chemical modifications.[4-8] PHAs, such as PHB and poly(3-hydroxybutyrate-*co*-3-hydroxyvalerate (PHBV), are brittle, which is related to their high crystalline degree and they may lack the superior mechanical properties required for biomedical and packaging applications. These properties are a consequence of PHA's chemical structure. Therefore, since these different types of PHA have various structural and physical chemical properties, they should be classified according to their properties and modified in order to be easy to use for target applications. Different approaches were explored for the production of PHA_{SCL} by the *Cupriavidus* genus. The main studies were: the utilization of noble *versus* waste carbon sources and the utilization of a limiting factor to trigger PHA production *versus* the operation under "nutrient-sufficient" conditions. Thus, Chen *et al.* (2011) observed that a smaller C–N ratio was more favorable for PHA accumulation in a culture of Gamma proteobacterium in 72 hours, whereas a higher C–N ratio was more favorable for PHA accumulation in longer cultures of up to 150 hours of cultivation.[9] Concerning activated sludge systems, Moralejo-Garate *et al.* (2013) have shown that the presence of ammonia during the PHA accumulation step was not damaging for PHA production.[10]

4-Hydroxybutyrate (4-HB) was produced by *Aeromonas hydrophila* 4AK4, *Escherichia coli* S17-1, or *Pseudomonas putida* KT2442 harboring 1,3-propanediol dehydrogenase gene dhaT and aldehyde dehydrogenase gene aldD from *P. putida* KT2442, which are capable of transforming 1,4-butanediol (1,4-BD) to 4HB. 4HB containing fermentation broth was used for the production of homopolymer poly-4-hydroxybutyrate [P(4HB)] and copolymer poly(3-hydroxybutyrate-*co*-4-hydroxybutyrate) [P(3HB-4HB)]. In this respect, attention has been given to producing PHAs bearing terminal double bonds in their side chains. The reason behind the preference for unsaturated PHAs can be found in the possibility to establish high-yield production processes.[11] Alkylic substrates containing double bonds are cheaper and generally exhibit less toxicity compared to substrates with reactive functional groups. At the same time, the unsaturated side chains of PHAs are susceptible to chemical modifications. Starting from alkene function, PHAs have been modified to exhibit functionalities *en route* towards new PHA-based biomaterials. The mechanical properties[12,13] of PHAs are directly correlated with their structure and crystallinity. An increase in the variety of side chains within one polymer chain of PHA$_{MCL}$ can modify its ability to crystallize and as a consequence there are some distinct differences in the crystallinity of PHA$_{MCL}$. Obtaining a low crystallinity is possibly done once the polymers have large and irregular pendant side groups attached. These groups inhibit the close packing of the polymeric chains in a regular three-dimensional fashion to form a crystalline array.

The physical and material properties of PHAs are greatly influenced by their monomer composition and chemical structure *i.e.* the length of the pendant groups that extend from the polymer backbone, the chemical nature of the pendant groups and the distance between the ester linkages in the polymer. The variety of bacterial PHA that can be directly produced by fermentation is extraordinary large with more than 150 different hydroxyalkanoic acids, and even mercaptoalkanoic acid, that are known constituents of these PHAs.[14] Depending on the subunit compositions and substrate specificities of the enzymes, PHA synthases are generally classified into four groups: class I, II, III, and IV.[15]

1.2 Recovery of Polyhydroxyalkanoates

The recovery system may affect the amount of product recovered, the convenience of the subsequent purification steps and the quality of the final product. Cell separation from the fermentation broth is the preliminary step of the recovery method. In order to recover the PHA granules, it is necessary to rupture the bacterial cell and remove the protein layer that coats the PHA granules. Alternatively, the PHA has to be selectively dissolved in a suitable solvent. Generally, two methods are usually utilized for the recovery and purification of PHAs from cell biomass, which include PHA solubilization or non-polymer cellular material (NPCM) dissolution. The majority of the PHA recovery method is performed using a solvent extraction process mainly by chloroform and methanol. Modifying the cell wall's permeability and then PHA dissolution in the solvent are the mechanisms for PHA extraction.

The separation of PHA from the solvent is carried out using solvent evaporation or polymer precipitation in a non-solvent material.[16] PHA is very viscous and the removal of cell debris is difficult. Without considering solvent recycling, the large amount of solvent required for PHA extraction is costly.[17] Consequently, PHA solubilizing into an immiscible solvent in water was explored at high temperature (above 120 °C). Then, cold water is added to extract the PHA, although the solvent can be recycled many times before being distilled.[18]

A new PHA recovery process was developed to obtain the benefits of both digestion methods using sodium hypochlorite and chloroform in a solvent extraction. The combined method creates three separate phases, which include hypochlorite solution at the upper phase, NPCM and undisrupted cells at the middle phase and a chloroform phase containing PHA. The polymer is then recovered by precipitation in a non-solvent and filtration. The molecular weight reduction due to polymer degradation is significantly reduced using this process.[19]

A few mechanical methods have also been developed to supplement these systems or as independent systems, which are widely used to recover intracellular PHA.[20] This field involves either solid shear (*e.g.* bead milling, extrusion of frozen cells) or liquid shear (*e.g.* high pressure homogenization). Combinations of methods, such as chemical and physical processes, can sometimes produce acceptable results whilst one method alone fails.[21] It has been reported that chemical pretreatments increase the sensitivity of bacteria to disruption. They allow equal disruption to be obtained at lower operating pressures or fewer passes during the physical process. The success of PHA as a viable option to petrochemical-based plastics will depend upon the design and performance of efficient and selective means of PHA production and recovery.[22] Thus, further investigations on mixed cultures, recombinant microbial strains, cheap carbon substrates and efficient fermentations has allowed the production of significant quantities of PHAs, which can significantly decrease the production cost. A commercial recovery system with a simple, efficient and economical procedure will probably focus on a non-solvent extraction-based recovery amongst a variety of PHA recovery methods. In addition, the tolerance of the final product to the conditions employed is an important criterion for the selection of a PHA recovery process and the PHA properties have to be considered for the development of downstream processes. If the PHA molecular weight is too low, the transition temperatures and the mechanical properties will usually decrease, which is not suitable for any useful commercial applications. Hence, the challenge in the recovery process should be the maintenance of the original molecular weights, while not compromising the degree of purity for various applications.[23] However, severe degradation of polymer molecular weight was reported during PHA extraction of *C. necator* using a sodium hypochlorite treatment.[24] It is necessary to characterize and compare the extracted and non-extracted polymer properties to assess the feasibility of the developed recovery methods on PHA extraction and address the possible market demand and intended applications. The selection of suitable PHA extraction methods depends on several process parameters such as concentration of chemicals, reaction time, recovery temperature and pH. Basically, the impact of process parameters on the effectiveness of PHA extraction procedures have been studied and proven, but there is a limitation of concrete data on the effect of external factors on PHA recovery.

1.3   Blends of Polyhydroxyalkanoates

Various PHA blends have been developed to improve the performance and to offset the high price of PHAs. The blending of PHAs will offer more scope to expand their range of applications. The P(3HB)/PLA blend is one of the most studied blends, which exhibits mechanical properties that are intermediate between the individual components. Although PLA and P(3HB) are biodegradable polymers synthesized from renewable resources, their potential applications are hampered due to brittleness and formation of very large spherulites.[25] Zhao and co-workers (2013) reported the preparation of a P(3HB-*co*-3HV)/PLA blend using a co-rotating twin-screw extruder. The melt mixing was carried out above the glass transition temperature (T_g) of amorphous polymer. The reason for using a twin-screw extruder was to ensure that all the specimens undergo the same thermal–mechanical history. Blending of PCL and P(3HB) offers a good option to improve the performance of both homopolymers.[26] They had prepared a blend of P(3HB)/PCL by melting the mixture in an internal mixer with compositions of PCL varying from 0 to 30 wt% to study the miscibility, morphology and physical–chemical properties of these systems.

Cellulose derivatives also have attracted much interest for their compatibility with P(3HB).[27,28] Ethyl-cellulose (EtC) is also a biomaterial like P(3HB) that is approved by the FDA (Food and Drug Administration) and is widely used as a blood coagulant, in coatings for pharmaceutical tablets and matrices for poorly soluble drugs. Zhang and co-workers (1997)[29] had investigated the miscibility, thermal behaviour and morphological structure of P(3HB) with ethyl cellulose (EtC) blends. A P(3HB)/starch blend was prepared either by a conventional solvent casting method or by melt processing methods, such as injection molding and compression molding after compounding. Two types of maize starch, Starch 1 (containing 70% amylose) and Starch 2 (containing 72% amylopectin), were blended with P(3HB) using a melt compounding method at a ratio of 70 : 30 wt% and characterised in terms of their morphology, structure, thermal, rheological and mechanical properties.[30] Ikejima and co-workers (1999)[31] had prepared P(3HB)/chitosan blend films in order to investigate the effect of deacetylation on the crystallisation behaviour of P(3HB). Chitosan is a copolysaccharide with a high degree of deacetylation. A solvent-casting method was employed to prepare the P(3HB)/chitosan blend films. P(3HB) and chitosan were dissolved separately in HFIP (1,1,1,3,3,3-hexafluoro-2-propanol) before blending.

Zembouai *et al.* (2013)[32] reported that the degradation of each polymer occurred separately and it was found that PLA was more thermally stable than P(3HB-*co*-3HV) copolymer. They also reported that the decomposition temperature, T_d, for all the blends was between PLA and the P(3HB-*co*-3HV) copolymer. Therefore, the thermal stability of the P(3HB-*co*-3HV)/PLA blends could be improved by increasing the amount of PLA. The T_m for the blend was found to be in between these two polymers. Miscibility of any polymer can be determined by evaluating the T_g. A single T_g indicates the miscibility of the polymer.[33] A study by Nanda *et al.* (2011)[34] reported one T_g which corresponded to PLA and the T_g for P(3HB-*co*-3HV) copolymer was poorly observed. They also reported that when the content of P(3HB-*co*-3HV) copolymer was increased, a reduction in T_g from 60 °C to 45 °C was observed. The same observation was reported by Richards *et al.* (2008)[33] and Modi *et al.* (2012).[35] Chee and co-workers (2002)[36] had performed viscometric studies on polymer-blend solutions of P(3HB) with PCL because it is a powerful method to assess the miscibility of the components in an amorphous state. Viscometric analysis demonstrated that P(3HB) was immiscible with PCl. Phase behaviour and crystallisation kinetics for the binary blend P(3HB)/cellulose propionate (CP) had been studied by Maekawa and co-workers (1999).[37] A strong dependence of the measure T_g on composition was detected at high levels of CP. The Flory–Fox equation is one of the best equations to describe the dependence of T_g on composition in miscible blend systems. Wang and co-workers (2003)[38] studied the miscibility, crystallisation behaviour, tensile properties and environmental biodegradability of P(3HB)/cellulose acetate butyrate (CAB) blends. Ismail *et al.* (2010)[39] studied the effect of starch content in a P(3HB) film matrix on its degree of swelling in water. Swelling in water and degradability are the most important characteristic for biodegradable materials. Polymer films were degraded by surface absorption of moisture and microorganisms.

1.4   Nanocomposites of Polyhydroxyalkanoates

Ten *et al.*[40] produced evenly exfoliated nanocomposites of PHBV/cellulose nanowhiskers, with a well-defined distribution of nanofillers within the polymer matrix. Zhijiang *et al.*[41] reported an improvement in the mechanical properties of a PHB nanocomposite made of bacterial cellulose nanofibrils that was prepared by a solution-casting method. In addition, they found the nanocomposite to show better biocompatibility and mechanical properties than pure PHB based on cell-adhesion analysis using Chinese hamster lung (CHL) fibroblast cells and stress strain tests, respectively. Ten *et al.*[42] studied the isothermal crystallization kinetics of poly(3-hydroxybutyrate-*co*-3-hydroxyvalerate) (PHBV) nanocomposites containing cellulose nanowhiskers (CNWs). The researchers studied the effects of the CNWs concentration and temperature on the crystallization rate and crystallinity of PHBV nanocomposites. Mook Choi *et al.*[43] for the first time described the production of an intercalated PHBHV-clay nanocomposite using melt extrusion in a

Brabender mixer at 165 °C, 50 rpm agitation rate for 15 min. The same melt extrusion was employed by Maiti *et al.*[44] to produce a well dispersed PHB/layered silicate nanocomposite. Zhang *et al.*[45] prepared nanocomposites of poly(3-hydroxybutyrate-*co*-3-hydroxyhexanoate) (PHBHHx)/layered silicates and PHBHHx/expanded graphite. For the nanofillers used in the two nanocomposites, at a lower nanofiller content the researchers observed an exfoliated morphology with good dispersion. Gorrasi *et al.*[46] obtained exfoliated nanocomposites of a poly-6-hydroxyhexanaote matrix through *in situ* intercalative ring opening polymerization of caprolactone with modified montmorillonite in the presence of a dibutyltin dimethoxide catalyst. Reports on the application of *in situ* intercalative polymerization for the production of bacterial PHA nanocomposites, especially those containing a medium-chain-length poly-3-hydroxyalkanoate matrix, are virtually non-existent.[47] Zhang *et al.*[29] suggested that the dispersion of the polymeric side chain into the nanofiller's layered spaces results in exfoliation. The characterization of the state of nanoparticle dispersion allows for the interpretation of the preceding morphologies, and the structural characterization relies heavily on techniques such as X-ray diffraction (XRD), wide angle X-ray diffraction (WAXD), simultaneous small angle X-ray scattering (SAXS) and electron microscopy (transmission, TEM, or scanning, SEM). Zhijiang *et al.*[47] employed the use of field emission scanning electron microscopy (FESEM) to qualitatively characterize a PHA/bacterial cellulose nanocomposite. Zhang *et al.*[48,49] reported the use of nanocomposite particles based on a PHA/folate ligand which carries doxorubicin as an anticancer therapy. The researchers reported that the membrane showed an excellent selectivity with relatively high permeation flux towards water compared to existing commercial membranes. Shape memory nanocomposite polymers have the ability to revert back to their original shape after being deformed thus expanding their applications to such areas as dry adhesion, microfluidics, biosensors and tissue engineering.[50] Ishida *et al.*[51] reported the synthesis of a shape memory polymer nanocomposite derived from poly(3-hydroxyoctanoate-*co*-3-hydroxyundecenoate) composited with nanofillers of silsesquioxane (POSS). A nanocomposite membrane based on a PHB-functionalized multi-walled carbon nanotubes/chitosan matrix was used to efficiently pervaporate a mixture of water and 1,4-dioxane without the use of an entrainer.[52] The researchers reported that the membrane showed an excellent selectivity with relatively high permeation flux towards water compared to existing commercial membranes.

1.5 Polyhydroxyalkanoate-based Multiphase Materials

Extrusion of PHA-based materials is in general linked with another processing step such as thermoforming, injection molding, fibre drawing, film blowing, bottle blowing or extrusion coating. PHA shows a low degradation temperature compared to its melting temperature. For instance, PHB homopolymer presents a narrow window for processing conditions. The PHB thermal[53–60] and thermo–mechanical[61,62] stabilities have been well described in the literature, demonstrating that the thermal degradation occurs in a one-step process, namely a random chain scission reaction. To improve such a drawback or to create new PHA properties, a great number of multiphase materials have been developed, mainly by mixing PHB or PHBV with others products such as plasticizers, fillers or other polymers. Many authors have noticed that PHA properties can evolve when plasticization occurs, *e.g.* with citrate ester (triacetin).[63–65]

Wang *et al.* tested different plasticizers—dioctyl phthalate, dioctyl sebacate, and acetyl tributyl citrate (ATBC)—with PHB.[66] The effects of biodegradable plasticizers on the thermal and mechanical properties of PHBV were studied by Choi and Park[67] using thermal and mechanical analyses. Soybean oil (SO), epoxidized soybean oil (ESO), dibutyl phthalate (DBP) and acetyl tributyl citrate (ATBC) were tested as plasticizing additives. Several blending studies have been performed on PHB or PHBV copolymers with a range of compounds that can help vary the crystallinity and degradation of the final blends.[68,69] The possibility of blending other polymers with PHA offers the possibility of not only overcoming the drawbacks of a small processing window and low impact resistance of PHAs but can also modify the crystallization tendency and biodegradation rates. The possibility of H-bonding or formation of donor–acceptor interactions between PHA and the other blend constituents helps improve blend miscibility and reduces the tendency of phase separation.

Applications of such PHA-based multilayers as commodities are primarily limited by PHA cost and until now by PHA availability, and thus attention is being focused on products with plastics constituting only a minor part of the product, such as paper coatings such as the plastic film moisture barrier in food or drink cartons and in sanitary napkins. The presence of cellulose fibres also increases the rate of PHBV crystallization, due to a nucleating effect, while thermal parameters, such as crystallinity content, remained unchanged. Studies on the crystallization behaviour of PHB/kenaf fibre biocomposites showed that the nucleation by kenaf fibres affected the crystallization kinetics of the PHB matrix.[70] Differences in the effect of cellulose fibres on the crystallization process have been attributed to the lignin content at the surface/interface of the cellulose fibre. The addition of cellulose fibres led to some improvement in tensile strength and stiffness, but the composites remained brittle.[71] At low content, the incorporation of cellulose fibres lowered the stiffness, however, higher amounts of cellulose fibres greatly improved the mechanical properties of PHB. For biocomposites based on cellulose fibres and PHB, the effect of fibre length, surface modification on the tensile and flexural properties has been investigated. Results on PHB reinforced with straw fibres have been published.[72] The structure–property relationships for PHA/OMMT nano-biocomposites were established and are in good agreement with the conclusions drawn in previously reported studies on synthetic polymer-based nanocomposites. Further attention was paid to the PHAs' degradation in nanocomposite systems since these polymers are very temperature sensitive.[73–75]

Thus, scientists were interested in other PHA-based nanocomposites filled with *e.g.* layered double hydroxides (LDH),[76,77] cellulose whiskers[78–80] and hydroxyapatite (HA),[81] the latter being used in particular for biomedical and tissue engineering applications. LDH structures are similar to layered silicate clays.

1.6   Modification of Polyhydroxyalkanoates

Chemical, physical and enzymatic approaches have been explored for polymer modifications, resulting in a uniquely transformed PHA endowed with a functionalized reactive group and/or enhanced properties such as thermal stability, elasticity, improved hydrophilicity and degradability. While chemical modification processes provide a large degree of freedom in controlling and designing the modified PHA in bulk quantities to suit a particular function, most often they have to contend with the drawback of toxic impurities that require difficult downstream processing. The structure of a PHA can be altered chemically to produce a modified polymer with predictable variation in molecular weight and functionality. For example, the hydrolytic rate of PHA to give an activated macromer that can accept a reactive functional group is said to depend on several factors such as the chemical nature or reactivity of the ester linkages between the monomers.[82] PHA modification by carboxylation is the addition of a carboxylic functional group to the polymeric macromer. Carboxylic groups incorporated into the polymer usually serve as functional binding sites for bioactive moieties such as probes for targeting proteins and hydrophilic components.[83]

Condensation reactions between carboxylic acids and amine groups were exploited to graft a modified PHA and linoleic acid onto chitosan.[84] Recently, Babinot *et al.*[85] used click ligation to esterify the pendant –COOH of carboxylated PHA with propargyl alcohol resulting in a clickable-alkyne group that was subsequently used to copolymerize a poly(ethylene glycol) (PEG) macromer onto the modified PHA. The properties of PHA and its copolymers have been reported to be modified by hydroxylation.[86–89] Normally, acid- or base-catalyzed reactions are used in the modification of PHA by hydroxylation in the presence of low molecular weight mono or diol compounds. Hydroxy-terminated PHA is important in block copolymerization. Methanolysis of PHA resulted in PHA methyl esters bearing monohydroxy-terminated groups. Halogenation of PHA is considered as an excellent method to diversify the polymer's functions and applications. Halogen atoms such as chlorine, bromine and fluorine were added to the olefinic bonds of unsaturated PHA through an addition reaction,[90] and to the saturated PHA *via* substitution reactions.

In another study, Arkin and Hazer[91] modified PHA-Cl into quaternary ammonium salts, thiosulfate moieties and phenyl derivatives. In addition, they cross-linked the modified PHA-Cl with benzene by electrophilic aromatic substitution using a Friedel–Crafts reaction. Mihara *et al.*[92] and Imamura *et al.*[93] filed an embodiment that detailed the procedures for PHA chemical modification by sulfanyl halogenation and the potential application of the modified PHA as a toner electrostatic charge controller in electrophotographic imaging. Another method to modify PHA is by graft copolymerization, which results in the formation of modified segmented copolymers with improved properties such as increased wettability and thermo–mechanical strength. Grafting reactions can be induced by either chemical, radiation or plasma discharge methods.[94,95] Chemical modification methods are sometimes aggressive and lead to reduced polymer molecular weight, unwanted side reaction(s) and toxic impurities. In some instances, a mild surface modification process is required without which the polymer may fail in its intended application(s). Irradiation of polymeric materials required no addition of polymer contaminants. Irradiations such as gamma-irradiation normally result in three-dimensional network structures with improved tensile strength. Several studies have demonstrated the cross-linking of unsaturated mcl-PHAs by gamma-irradiation.[96–98] The presence of olefinic bonds in PHA side chains provides an avenue for polymer modification by several irradiation processes. A highly cross-linked modified polymer was produced by irradiating unsaturated PHA obtained from tallow-grown *P. resinovorans* with 25–50 kilogray (kGy) of γ-irradiation. Ion implantation is another physical method employed in polymer surface modification. Its advantage over polymer modification methods is that it only modifies the polymer surface layer, without upsetting the bulk polymer's properties. Ion implantation has been successfully applied in several polymer modifications thereby expanding its applications.[99–103] Mirmohammadi *et al.*[104] compared the biocompatibility of a PHB surface upon treatment with O_2 and CO_2 plasma at 50 W discharge for 3 min, and found that O_2 plasma treated PHB showed much improvement. Ying *et al.*[105] evaluated the biocompatibility and biosorption characteristics of an electrospun scaffold of P3HB4HB through subcutaneous implantation of the fibers in rats. The researchers found a highly increased tissue response with increasing content of 4HB monomer.

1.7   Polyhydroxyalkanoates as Packaging Materials: Current Applications and Future Prospects

The most common PHA packaging resins are polyhydroxybutyrate (PHB) and its copolymer with polyhydroxyvalerate (P(HB-*co*-HV)). The potential of PHAs as biodegradable replacements for conventional bulk commodity plastic packaging while promoting sustainable development has long been recognised.[106] The potential of PHAs for truly biodegradable packaging was recognised in the 1980s with the commercial release of Biopol®, thermoplastic resins of P(3HB) with various copolymer loadings of (3HV), by Imperial Chemical Industries (ICI, now Zeneca). One of the prohibiting factors against the use of PHAs as a resin for packaging materials is that it is economically uncompetitive in the current market compared to fossil fuel sourced synthetic raw material (ff-polymers).[107,108] In the production of PHAs, the cost of the carbon substrate represents approximately 50% of the total production cost. In order to rival current synthetic polymers used in packaging such as polyethylene (PE), polypropylene (PP) or polystyrene (PS) for example, PHAs' physical and chemical properties need to be comparable. The optical properties of plastic packaging, certainly in the case of food packaging, offer a convenient, lightweight and flexible adaption of packaging technologies for the food industry, reducing the reliance upon glass and metallic canning. Transparency, a variety of packaging options such as shrink wrap, a modified atmosphere and printability allow plastic packaging to be tailored to the type of food to be contained. The thermal properties are a vital consideration when selecting a polymer for packaging. Fortunately, PHAs provide (through diversity of structure and chemistry) a wide range of thermal properties for selection to suit packaging needs. Melting temperatures (T_m) from 60 to 177 °C, glass transition temperatures (T_g) from –50 to 4 °C and thermal degradation temperatures at highs of 256 to 277 °C are all within the range of the PHAs currently being produced.[109] Carboxyl-terminated butadiene acrylonitrile rubber (CTBA) and polyvinyl pyrrolidone (PVP) have been added to PHB in an effort to modify its thermal processing. Hong *et al.* report a significant modification of PHB crystallisation rate, crystallinity, melting temperature and thermal stability with the addition of only 1% (w/w) of these additives.[110] When compared to the other typical ff-polymers of polyethylene (PE) and polystyrene (PS) used in packaging, the oxygen and water barrier properties of commercial packaging PHA resins are considered to be naturally of a superior level. The vapour pressure exerted by these small aroma bearing molecules provides an added challenge for the application of biopolymers in packaging. A packaging's vapour barrier properties relates to its ability to prevent water vapour from crossing the polymer packaging boundary. Several factors including mechanical, morphology and crystallinity can play a substantial role in determining a packaging's vapour barrier properties. One particular strategy of improving PHA packaging barrier properties would be to develop suitable nanocomposites. In particular, nanocomposites incorporating nanoclays of montmorillonite and kaolinite clays could also substantially improve the mechanical strength and thermal stability as well as the gas barrier properties. In a similar strategy as that used to improve the thermal properties of polymers, PHAs have been used as additives to improve the barrier properties of conventional, synthetic chemicals. For example, addition of P(3HB) to polyvinyl alcohol (PVOH) can lead to significant improvement in its barrier properties. The herbicide product was successful delivered over a period of time with gradual degradation of the PHA packaging and was successful in controlling the growth of creeping bentgrass, *Agrostis stolonifera*.[111] Notably, adjustment of the content of the PHA casing enabled a level of control over degradation rate and product release.

1.8 Packaging Applications of Polyhydroxyalkanoates

According to Robertson (2010),[112] the main functions of packaging are to contain, to protect, to be convenient, to communicate and to sell the product. The basic functions of a package are to contain a certain amount of food, unitizing the product and facilitating its transportation, storage, sale and use. Bucci *et al.* (2005)[113] reported that PHB can be used in injection molding processes for the manufacture of food packaging, with the same equipment used for PP packaging injection. However, the process conditions should be adjusted according to polymer characteristics. The authors found a notable difference between PHB and PP bottles in relation to their performance in dynamic compression resistance and a drop test, in that PHB is as hard a material as PP, but is less flexible. PHB performance was better at higher temperatures. The physical, dimensional, mechanical and sensory tests showed that PHB can replace PP containers for food products with high fat content (mayonnaise, margarine and cream cheese), including storage in freezers and heating in microwave ovens. Likewise, Muizniece-Brasava and Dukalska (2006)[114] reported that PHB materials are suitable for sour cream storage. To improve flexibility for potential packaging applications, PHB is synthesized with various co-polymers, such as poly (3-hydroxyvalerate) (HV), leading to a decrease of the glass transitions and melting temperatures. In addition, the HV broadens the processing window since there is improved melt stability at lower processing temperatures (Modi *et al.*, 2010).[115] Fabra *et al.* (2013)[116] created an innovative way to develop renewable biopolyester microbial-based multilayer structures with enhanced barrier performance, which is of significant interest for food packaging applications. The researchers developed multilayer structures based on polyhydroxybutyrate-*co*-valerate with a valerate content of 12% (PHBV12) containing a high barrier interlayer of zein electrospun nanofibers. The incorporation of 3-hydroxyvalerate (HV) in PHB, resulting in PHBV, has increased impact strength, elongation modulus, tensile strength and decreased Young's modulus, making the film more flexible and more resistant (Shen *et al.*, 2009).[117] The price is very high but PHBV degrades between five and six weeks in a microbiologically active environment, resulting in water and carbon dioxide in aerobic conditions. In an anaerobic environment, degradation is faster, producing methane (Siracusa *et al.*, 2008).[118]

A mixture of PHBV with PLA had a positive effect on the elasticity modulus, elongation at break and flexural strength for different blends. However, tensile strength did not improve in any of them. In the same way, Zhang *et al.* (1996)[119] reported improved mechanical properties for blends of PHB/PLA compared with the common PHB. In addition, PVA (poly-vinylacetate) grafted on PIP (poly-*cis*-1,4-isoprene) and mixed with PHB had better tensile properties and impact strength than PHB/PIP blends, which were immiscible (Yoon *et al.*, 1999).[120] Combined with synthetic plastics or starch, PHAs make excellent packaging films (Tharanathan, 2003).[121] In another study conducted by Shen *et al.* (2009),[122] thermoplastic starch (TPS) blended with PHA had a positive effect on the barrier and hydrolytic properties and UV stability of a starch-based film. With this blend, it was possible to reduce the processing temperature, resulting in less degradation of the starch.

References

1. J. M. Luengo, B. García, A. Sandoval, G. Naharro and E. a. R. Olivera, *Curr. Opin. Microbiol.*, 2003, **6**, 251–260.
2. R. Rai, T. Keshavarz, J. A. Roether, A. R. Boccaccini and I. Roy, *Mater. Sci. Eng., R*, 2011, **72**, 29–47.
3. D. B. Hazer, E. Kılıçay and B. Hazer, *Mater. Sci. Eng., C*, 2012, **32**, 637–647.
4. E. N. Pederson, C. W. J. McChalicher and F. Srienc, *Biomacromolecules*, 2006, **7**, 1904–1911.
5. L. Tripathi, L.-P. Wu, M. Dechuan, J. Chen, Q. Wu and G.-Q. Chen, *Bioresour. Technol.*, 2013, **142**, 225–231.
6. L. Zhang, Z.-Y. Shi, Q. Wu and G.-Q. Chen, *Appl. Microbiol. Biotechnol.*, 2009, **84**, 909–916.
7. I. K. Kang, S. H. Choi, D. S. Shin and S. C. Yoon, *Int. J. Biol. Macromol.*, 2001, **28**, 205–212.
8. B. Hazer and A. Steinbüchel, *Appl. Microbiol. Biotechnol.*, 2007, **74**, 1–12.
9. Z. Chen, Y. Li, Q. Wen and H. Zhang, *Chemosphere*, 2011, **82**, 1209–1213.
10. H. Moralejo-Garate, T. Palmeiro-Sanchez, R. Kleerbezem, A. Mosquera-Corral, J. L. Campos and M. C. M. van Loosdrecht, *Biotechnol. Bioeng.*, 2013, **110**, 3148–3155.
11. P. Höfer, P. Vermette and D. Groleau, *Biochem. Eng. J.*, 2011, **54**, 26–33.
12. K. Sudesh, H. Abe and Y. Doi, *Prog. Polym. Sci.*, 2000, **25**, 1503–1555.
13. P. Hoefer, *Front. Biosci.*, 2010, **15**, 93–121.
14. B. Hazer and A. Steinbüchel, *Appl. Microbiol. Biotechnol.*, 2007, **74**, 1–12.
15. S. J. Park, T. W. Kim, M. K. Kim, S. Y. Lee and S.-C. Lim, *Biotechnol. Adv.*, 2012, **30**, 1196–1206.
16. D. Byrom, *Polyhydroxyalkanoates*, Hanser, Munich, 1994, 5–33.
17. Y. H. Chen, M. L. Wu and W. M. Fu, *J. Physiol.*, 1998, **507**, 41–53.
18. J. M. Liddell, *Process for the recovery of polyhydroxyalkanoic acid*, United States Patent, US 5894062, 1999.
19. S. K. Hahn, Y. K. Chang, B. S. Kim and H. N. Chang, *Biotechnol. Bioeng.*, 1994, **44**, 256.
20. M. Tamer and M. Moo-Young, *Bioprocess Biosyst. Eng.*, 1998, **19**, 459–468.
21. T. R. Hopkins, *J. Bioprocess Technol.*, 1991, **12**, 57–83.

22. Y. Kathiraser, M. K. Aroua, K. B. Ramachandran and I. K. P. Tan, *J. Chem. Technol. Biotechnol.*, 2007, **82**, 847–855.
23. S. K. Hahn, Y. K. Chang, B. S. Kim and H. N. Chang, *Biotechnol. Bioeng.*, 1994, **44**, 256–261.
24. B. Kunasundari and K. Sudesh, *eXPRESS Polym. Lett.*, 2011, **5**, 620–634.
25. M. Avella, E. Martuscelli and M. Raimo, *J. Mater. Sci.*, 2000, **35**, 523.
26. H. Zhao, Z. Cui, X. Sun, L. H. Turng and X. Peng, *Ind. Eng. Chem. Res.*, 2013, **52**, 2569.
27. M. C. M. Antunes and M. I. Felisberti, *Polim.: Cienc. Tecnol.*, 2005, **15**, 134–138.
28. L. Yu, K. Dean and L. Li, *Prog. Polym. Sci.*, 2006, **31**, 576–602.
29. L. Zhang, X. Deng and Z. Huang, *Polymer*, 1997, **38**, 5379–5387.
30. M. Zhang and N. L. Thomas, *J. Appl. Polym. Sci.*, 2010, **116**, 688–694.
31. T. Ikejima, K. Yagi and Y. Inoue, *Macromol. Chem. Phys.*, 1999, **200**, 413–421.
32. I. Zembouai, M. Kaci, S. Bruzaud, A. Benhamida, Y. M. Corre and Y. Grohens, *Polym. Test.*, 2013, **32**, 842–851.
33. E. Richards, R. Rizvi, A. Chow and H. Naguib, *J. Polym. Environ.*, 2008, **16**, 258–266.
34. M. R. Nanda, M. Misra and A. K. Mohanty, *Macromol. Mater. Eng.*, 2011, **296**, 719–728.
35. S. Modi, K. Koelling and Y. Vodovotz, *J. Appl. Polym. Sci.*, 2012, **124**, 3074–3081.
36. M. J. K. Chee, J. Ismail, C. Kummerlöwe and H. W. Kammer, *Polymer*, 2002, **43**, 1235–1239.
37. M. Maekawa, R. Pearce, R. H. Marchessault and R. S. J. Manley, *Polymer*, 1999, **40**, 1501–1505.
38. T. Wang, G. Cheng, S. Ma, Z. Cai and L. Zhang, *J. Appl. Polym. Sci.*, 2003, **89**, 2116–2122.
39. A. M. Ismail and M. A. B. Gamal, *J. Appl. Polym. Sci.*, 2010, **115**, 2813–2819.
40. E. Ten, J. Turtle, D. Bahr, L. Jiang and M. Wolcott, *Polymer*, 2010, **51**, 2652–2660.
41. C. Zhijiang, Y. Guang and J. Kim, *Curr. Appl. Phys.*, 2011, **11**, 247–249.
42. E. Ten, D. F. Bahr, B. Li, L. Jiang and M. P. Wolcott, *Ind. Eng. Chem. Res.*, 2012, **51**, 2941–2951.
43. W. Mook Choi, T. Wan Kim, O. Ok Park, Y. Keun Chang and J. Woo Lee, *J. Appl. Polym. Sci.*, 2003, **90**, 525–529.
44. P. Maiti, C. A. Batt and E. P. Giannelis, *Biomacromolecules*, 2007, **8**, 3393–3400.
45. X. Zhang, G. Lin, R. Abou Hussein, W. M. Allen, I. Noda and J. E. Mark, *J. Macromol. Sci., Part A*, 2008, **45**, 431–439.
46. G. Gorrasi, M. Tortora, V. Vittoria, E. Pollet, B. Lepoittevin, M. Alexandre and P. Dubois, *Polymer*, 2003, **44**, 2271–2279.
47. C. Zhijiang, H. Chengwei and Y. Guang, *Carbohydr. Polym.*, 2012, **87**, 1073–1080.

48. X. Zhang, G. Lin, R. Abou-Hussein, M. K. Hassan, I. Noda and J. E. Mark, *Eur. Polym. J.*, 2007, **43**, 3128–3135.

49. C. Zhang, L. Zhao, Y. Dong, X. Zhang, J. Lin and Z. Chen, *Eur. J. Pharm. Biopharm.*, 2010, **76**, 10–16.

50. Y. T. Ong, A. L. Ahmad, S. H. S. Zein, K. Sudesh and S. H. Tan, *Sep. Purif. Technol.*, 2011, **76**, 419–427.

51. K. Ishida, R. Hortensius, X. Luo and P. T. Mather, *J. Polym. Sci., Part B: Polym. Phys.*, 2012, **50**, 387–393.

52. Y. T. Ong, A. L. Ahmad, S. H. S. Zein, K. Sudesh and S. H. Tan, *Sep. Purif. Technol.*, 2011, **76**, 419–427.

53. N. Grassie, E. J. Murray and P. A. Holmes, *Polym. Degrad. Stab.*, 1984, **6**, 47.

54. N. Grassie, E. J. Murray and P. A. Holmes, *Polym. Degrad. Stab.*, 1984, **6**, 95.

55. N. Grassie, E. J. Murray and P. A. Holmes, *Polym. Degrad. Stab.*, 1984, **6**, 127.

56. M. Kunioka and Y. Doi, *Macromolecules*, 1990, **23**, 1933.

57. Y. Aoyagi, K. Yamashita and K. Doi, *Polym. Degrad. Stab.*, 2002, **76**, 53.

58. S.-D. Li, J. D. He, P. H. Yu and M. K. Cheung, *J. Appl. Polym. Sci.*, 2003, **89**, 1530.

59. H. Abe, *Macromol. Biosci.*, 2006, **6**, 469.

60. F. Carrasco, D. Dionisi, A. Martinelli and M. Majone, *J. Appl. Polym. Sci.*, 2006, **100**, 2111.

61. D. H. Melik and L. A. Schechtman, *Polym. Eng. Sci.*, 1995, **35**, 1795.

62. R. Renstad, S. Karlsoon and A.-C. Albertsson, *Polym. Degrad. Stab*, 1997, **57**, 331.

63. M. A. Kotnis, G. S. O'Brien and J. L. Willett, *J. Environ. Polym. Degrad.*, 1995, **3**, 97.

64. R. L. Shogren, *J. Environ. Polym. Degrad.*, 1995, **3**, 75.

65. R. C. Baltieri, L. H. I. Mei and J. Bartoli, *Macromol. Symp.*, 2003, **197**, 33.

66. L. Wang, W. Zhu, X. Wang, X. Chen, G.-Q. Chen and K. Xu, *J. Appl. Polym. Sci.*, 2008, **107**, 166.

67. J. S. Choi and W. H. Park, *Polym. Test.*, 2004, **23**, 455.

68. J.-C. Huang, A. S. Shetty and M.-S. Wang, *Adv. Polym. Technol.*, 1990, **10**, 23.

69. H. Verhoogt, B. A. Ramsay and B. D. Favis, *Polymer*, 1994, **35**, 5155.

70. A. K. Mohanty, M. Misra and G. Hinrichsen, *Macromol. Mater. Eng.*, 2000, **276–277**, 1–24.

71. R. A. Shanks, A. Hodzic and S. Wong, *J. Appl. Polym. Sci.*, 2004, **91**, 2114.

72. P. Gatenholm, J. Kubat and A. Mathiasson, *J. Appl. Polym. Sci.*, 1992, **45**, 1667.

73. P. Bordes, E. Pollet and L. Averous, *Prog. Polym. Sci. (Oxford)*, 2009, **34**, 125.

74. L. Cabedo, D. Plackett, E. Gimenez and J. M. Lagaron, *J. Appl. Polym. Sci.*, 2009, **112**, 3669.

75. P. Bordes, E. Hablot, E. Pollet and L. Averous, *Polym. Degrad. Stab.*, 2009, **94**, 789.

76. E. Hablot, P. Bordes, E. Pollet and L. Averous, *Polym. Degrad. Stab.*, 2008, **93**, 413.

77. T. M. Wu, S.-F. Hsu and C.-S. Liao, *J. Polym. Sci., Part B: Polym. Phys.*, 2006, **44**, 3337.

78. S.-F. Hsu, T. M. Wu and C.-S. Liao, *J. Polym. Sci., Part B: Polym. Phys.*, 2007, **45**, 995.

79. D. Dubief, E. Samain and A. Dufresne, *Macromolecules*, 1999, **32**, 5765.

80. A. Dufresne, M. B. Kellerhals and B. Witholt, *Macromolecules*, 1999, **32**, 7396.

81. A. Dufresne, *Compos. Interfaces*, 2000, **7**, 53.

82. D. P. Martin, F. Skraly and S. F. Williams, Google Patents, 2003.

83. D. Kai and X. J. Loh, *ACS Sustainable Chem. Eng.*, 2013, **7**(10).

84. H. Arslan, B. Hazer and S. C. Yoon, *J. Appl. Polym. Sci.*, 2007, **103**, 81–89.

85. J. Babinot, E. Renard and V. Langlois, *Macromol. Chem. Phys.*, 2011, **212**, 278–285.

86. L. Zhou, Z. Chen, W. Chi, X. Yang, W. Wang and B. Zhang, *Biomaterials*, 2012, **33**, 2334–2344.

87. M. Kwiecień, G. Adamus and M. Kowalczuk, *Biomacromolecules*, 2013, **14**, 1181–1188.

88. D. Hu, A.-L. Chung, L.-P. Wu, X. Zhang, Q. Wu, J.-C. Chen and G.-Q. Chen, *Biomacromolecules*, 2011, **12**, 3166–3173.

89. A. P. Andrade, B. Witholt, R. Hany, T. Egli and Z. Li, *Macromolecules*, 2001, **35**, 684–689.

90. A. H. Arkin, B. Hazer and M. Borcakli, *Macromolecules*, 2000, **33**, 3219–3223.

91. A. H. Arkin and B. Hazer, *Biomacromolecules*, 2002, **3**, 1327–1335.

92. C. Mihara, T. Yano, S. Kozaki, T. Honma, T. Kenmoku, T. Fukui and A. Kusakari, Google Patents, 2008.

93. T. Imamura, E. Sugawa, T. Yano, T. Nomoto, T. Suzuki, T. Honma, T. Kenmoku and T. Fukui, Google Patents, 2008.

94. H. W. K. Do Young Kim, M. G. Chung and Y. H. Rhee, *J. Microbiol.*, 2007, 87–97.

95. S. Nguyen, *Can. J. Chem.*, 2008, **86**, 570–578.

96. R. D. Ashby, A.-M. Cromwick and T. A. Foglia, *Int. J. Biol. Macromol.*, 1998, **23**, 61–72.

97. R. D. Ashby, T. A. Foglia, C.-K. Liu and J. W. Hampson, *Biotechnol. Lett.*, 1998, **20**, 1047–1052.

98. A. Dufresne, L. Reche, R. H. Marchessault and M. Lacroix, *Int. J. Biol. Macromol.*, 2001, **29**, 73–82.

99. D. M. Zhang, F. Z. Cui, Z. S. Luo, Y. B. Lin, K. Zhao and G. Q. Chen, *Surf. Coat. Technol.*, 2000, **131**, 350–354.

100. N. J. Nosworthy, A. Kondyurin, M. M. M. Bilek and D. R. McKenzie, *Enzyme Microb. Technol.*, 2014, **54**, 20–24.

101. A. Belbah, A. Mkaddem, N. Ladaci, N. Mebarki and M. El Mansori, *Mater. Des.*, 2014, **53**, 202–208.

102. X.-Y. Chen, X.-F. Zhang, Y. Zhu, J. Zhang and P. Hu, *Polym. J.*, 2003, **35**, 148–154.

103. M. Manso, A. Valsesia, M. Lejeune, D. Gilliland, G. Ceccone and F. Rossi, *Acta Biomater.*, 2005, **1**, 431–440.

104. S. A. Mirmohammadi, M. T. Khorasani, H. Mirzadeh and S. Irani, *Polym.–Plast. Technol. Eng.*, 2012, **51**, 1319–1326.

105. T. H. Ying, D. Ishii, A. Mahara, S. Murakami, T. Yamaoka, K. Sudesh, R. Samian, M. Fujita, M. Maeda and T. Iwata, *Biomaterials*, 2008, **29**, 1307–1317.

106. G. Braunegg, G. Lefebvre and K. F. Genser, *J. Biotechnol.*, 1998, **65**, 127.

107. M. S. Sreekanth, S. V. N. Vijayendra, G. J. Joshi and T. R. Shamala, *J. Food Sci. Technol.*, 2013, **50**, 404.

108. Q. Chen and L. H. Zhang, *Appl. Mech. Mater.*, 2014, **448**, 160.

109. G.-Q. Chen and M. K. Patel, *Chem. Rev.*, 2012, **112**, 2082.

110. S.-H. Hong, T.-S. Gau and S.-C. Huang, *J. Thermal Anal. Calorim.*, 2011, **103**, 967.

111. V. Prudnikova, A. N. Boyandin, G. S. Kalacheva and A. J. Sinskey, *J. Polym. Environ.*, 2013, **21**, 675.

112. G. L. Robertson, *Food Packaging, Principles and Practice*, CRC Press, Boca Raton, 2nd edn, 2006, p. 550.

113. D. Z. Bucci, L. B. B. Tavares and I. Sell, *Polym. Test.*, 2005, **24**, 564–571.

114. S. Muizniece-Brasava and L. Dukalska, *Proceedings of the Latvia University of Agriculture*, 2006, **16**(311), 79–87.

115. S. J. Modi, Thesis: *Assessing the Feasibility of Poly-(3-Hydroxybutyrate-co-3-Valerate) (PHBV) and Poly-(Lactic Acid) for Potential Food Packaging Applications*, 2010.

116. M. J. Fabra, A. Lopez-Rubio and J. M. Lagaron, *Food Hydrocolloids*, 2013, **32**, 106–114.

117. Z. Shen, G. P. Simon and Y. B. Cheng, *Polymer*, 2002, **43**, 4251–4260.

118. V. Siracusa, P. Rocculi, S. Romani and M. Dalla Rosa, *Trends Food Sci. Technol.*, 2008, **19**(12), 634–643.

119. L. Zhang, C. Xiong and X. Deng, *Polymer*, 1996, **37**(2), 235–241.

120. J. S. Yoon, W. S. Lee, H. J. Jin, I. J. Chin, M. N. Kim and J. H. Go, *Eur. Polym. J.*, 1999, **35**(5), 781–788.

121. R. N. Tharanathan, *Trends Food Sci. Technol.*, 2003, **4**(3), 71–78.

122. L. Shen, J. Haufe and M. K. Patel, Product overview and market projection of emerging bio-based plastics, 2009.

CHAPTER 2

Polyhydroxyalkanoates: Structure, Properties and Sources

NATHALIE BEREZINA*[a] AND SILVIA MARIA MARTELLI[b]

[a] Materia Nova, Rue des Foudriers 1, 7822, Ghislenghien, Belgium;
[b] Faculty of Engineering, Federal University of Grande Dourados, Dourados, Brazil
*Email: nathalie.berezina@materianova.be

2.1 Introduction

Polyhydroxyalkanoates (PHAs) are biopolyesters that generally consist of (R)-3-hydroxy fatty acids, with various side chains (Figure 2.1). Other fatty acids have the hydroxyl group at the 4- or 5-position. The pendant group (R) varies from C1 to C13 and is saturated or unsaturated or contains a substituent. More than 150 kinds of hydroxycarboxylic acids have been identified as PHA monomers.[1] Regarding the length of the side chain, there are three types of PHA: (1) short chain length hydroxyalkanoic acids (PHA$_{SCL}$) with an alkyl side chain that are produced by *Ralstonia eutropha* and many other bacteria. PHA$_{SCL}$ contains 3–5 carbon atoms, for example poly-3-hydroxybutyrate (P3HB), poly-4-hydroxybutyrate (P4HB); (2) medium chain length hydroxyalkanoic acids (PHA$_{MCL}$) with an alkyl side chain that are produced by *Pseudomonas oleovorans* and other *Pseudomonas sensu stricto*. PHA$_{MCL}$ contain 6–14 carbon atoms and (3) long chain length (PHA$_{LCL}$) obtained from long chain fatty acids, which contain more than 14 carbon atoms.[2]

RSC Green Chemistry No. 30
Polyhydroxyalkanoate (PHA) Based Blends, Composites and Nanocomposites
Edited by Ipsita Roy and Visakh P M
© The Royal Society of Chemistry 2015
Published by the Royal Society of Chemistry, www.rsc.org

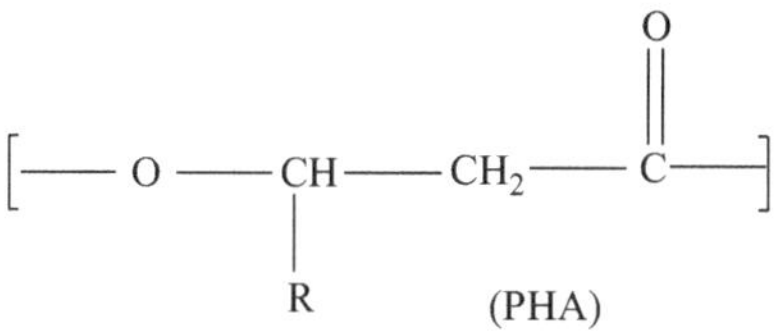

$$\left[\!\!-\!\!-\!O\!-\!\!-\!CH\!-\!\!-\!CH_2\!-\!\!-\!\overset{\overset{\textstyle O}{\|}}{C}\!-\!\!-\right]$$

$$\underset{R}{\big|}\qquad\text{(PHA)}$$

R = methyl or ethyl (short chain length)

R = propyl, butyl, pentyl, hexyl or heptyl (medium chain length)

R = more than 14 carbons per repeating unit (long chain length)

Figure 2.1 General classification of bacterial polyesters according to their chemical structure.

The large variability in the chemical structure and material properties of PHAs is due to the low substrate specificity of PHA synthases and the subsequent modifications by chemical reactions. The monomer composition, macromolecular structure and physical chemical properties of PHA vary depending on the producer organism as well as the carbon source used for the growth.[3–5] Their properties span a wide range, including materials that resemble polypropylene and others that are elastomeric.

The monomer composition of PHA has a considerable effect on its physical properties. For instance, in the case of PHA$_{MCL}$, the combination of T_g values below room temperature and a low degree of crystallinity imparts elastomeric behavior to these polymers. The PHA structure can effectively be controlled by adjusting the carbon substrates to achieve the desired monomer content, by engineering metabolic pathways in the hosts or by feeding the culture with carbon substrates containing functional side chains that in a second step can undergo chemical modifications.[6–10] Their different properties arise chemically, either from the length of the pendant groups which extend from the polymer backbones, or from the distance between the ester linkages in the polymer backbones. Typically, PHAs with short pendant groups are hard crystalline materials, whereas PHAs with longer pendant groups are elastomeric.[11,12]

PHAs such as PHB and PHBV are brittle, which is related to their high crystalline degree and they may lack the superior mechanical properties required for biomedical and packaging applications. The physical properties of P(3HB-co-4HB) and P(3HB-co-3HP) show some improved flexibility as biodegradable thermoplastics compared to those of PHB, as the 4HB and 3HP monomer fractions in the copolymer increase, respectively.[13,14] On the other hand, PHA$_{MCL}$ may be elastomeric but have very low mechanical strength.[10] These properties are a consequence of PHAs' chemical structure. Therefore, since these different types of PHAs have various structural and physical chemical properties, they should be classified according to their properties and modified in order to be easy to use for target applications.

In this chapter, recent developments in the field of PHA carbon sources, their properties and relations between sources, structure and properties are discussed.

2.2   Sources of PHAs

The biosynthesis of PHA_{SCL} and PHA_{MCL} is performed following different metabolic pathways (Figure 2.2).[15–17] Thus, PHA_{SCL} are synthesized in three main steps starting with acetyl-CoA, whereas PHA_{MCL} can be synthesized by two different pathways. The first pathway consists of the β-oxidation of fatty acids prior to their incorporation in the polyester chain, the second consists of the *de novo* synthesis of fatty acids.[15]

Several enzymes are involved in these metabolic pathways. In the case of PHA_{SCL}, the genes encoding PhaA, PhaB and PhaC are often co-localized and organized in an operon, whereas in the case of PHA_{MCL}, genes encoding PhaG and PhaJ are not co-localized, but co-regulated with PhaC.[15]

These different metabolic pathways show the importance of different carbon sources. Indeed, different carbon sources can be used for the production of PHA_{SCL}, whereas the choice of the carbon source deeply influences the structure of the resulting PHA_{MCL}, therefore this category of PHAs is discussed in the section on structure.

Several microorganisms are able to produce PHA_{SCL}, however, the *Cupriavidus* genus is by far the most studied for this purpose. Numerous reasons justify this choice: the genus is rather robust, accepts different

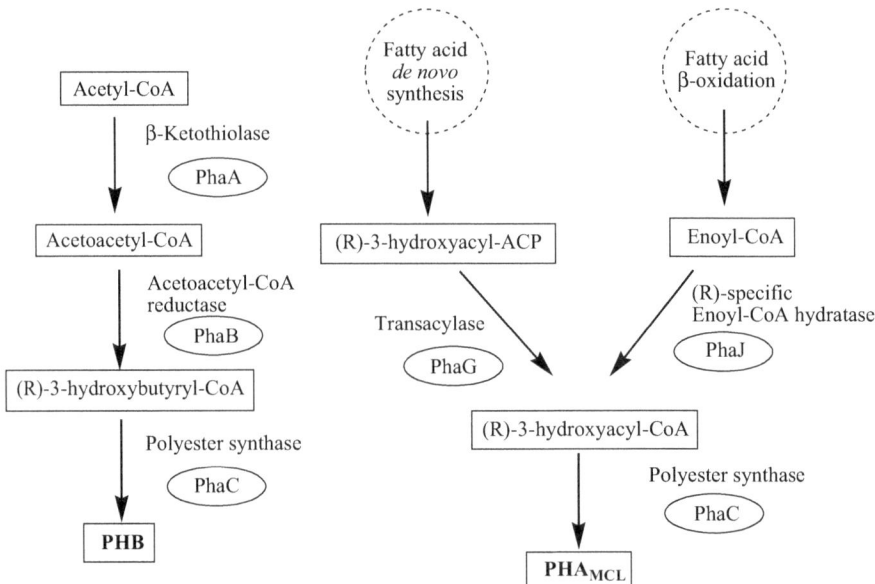

Figure 2.2   Biosynthetic pathways for PHA synthesis.

carbon sources (noble as well as wastes) and, generally, shows good productivity (Table 2.1).

Different approaches were explored for the production of PHA$_{SCL}$ by the *Cupriavidus* genus. The main approaches are: the utilization of noble *versus* waste carbon sources and the utilization of a limiting factor to trigger PHA production *versus* operation under "nutrient-sufficient" conditions.

Table 2.1 Sources of PHA for *Cupriavidus* genus.

Substrate	Limiting nutrient	PHA content (%)	Yield (g g^{-1})	Productivity (g L^{-1} h^{-1})	Ref.
Glucose	N[a]	76	0.3	2.42	18
Glucose	N[a]	76	n.d.	2.03	19
Glucose	P[a]	80	0.38	3.14	20
Glucose	P	50	n.d.	3.1	27
Glucose for growth butyric acid for PHA accumulation	P[a]	n.d.	0.82	0.57	21
Glucose + glutamate for growth glucose for PHA accumulation	-/-	>70	n.d.	>0.5	25
Glucose for growth peptone – for PHA accumulation	-/-	78	n.d.	1.2	24
Glutamate for growth no substrate for accumulation	-/-[a]	34	n.d.	n.d.	23
LB for growth no substrate for accumulation	-/-[a]	34	n.d.	n.d.	23
Glycerol	N	<60	0.16	n.d.	116
Glycerol	N	62	0.36	1.5	117
Mixture glucose + glycerol	N	70[b]	0.35[b]	n.d.	118
Waste glycerol	N	63	n.d.	1.36	19
Waste glycerol	N	50	0.34	1.1	29
Peptone + meat + yeast extract for growth OPF – for PHA accumulation	-/-	32	0.2	n.d.	119,120
LB – for growth ice cream residue – for PHA accumulation	-/-	87	n.d.	0.02	121
Acetic + butyric acids for growth digestable liquor – for PHA accumulation	-/-[a]	90	0.48	n.d.	22
Waste frying oil	-/-	88	0.64	n.d.	122
Meat extract + peptone for growth soybean oil – for PHA accumulation	-/-	81	0.85	2.5	26

n.d.: not determined;
[a]The influence of the limiting factor was a specific object of the study and is discussed in the chapter text;
[b]Only theoretical values, no practical experiments performed.

Among the different limiting factors, the most popular, until now, are nitrogen and phosphorus (Table 2.1). It is, however, worth noting, that the use of the limiting nutrient is more characteristic for growth on noble carbon sources, *i.e.* sugars, or on glycerol, either pure or waste. In cases when the waste carbon source is used, most of the time, no limiting factor is applied.

Some studies were specifically oriented towards evaluating the importance of the limiting factor. Thus, Kim *et al.* (1994) showed that *C. necator* performs better with a delayed nitrogen limitation.[18] More recently, Mozumber *et al.* (2013) established that the nitrogen limitation can, of course, be considered as a sufficient condition for the production of PHA by *C. necator*, but in no case can this be considered a necessary condition to trigger such a production.[19]

Concerning the phosphorus limitation, Ryu *et al.* (1997) have shown that the enhancement of its concentration improves PHA production.[20] The main argument in favor of the use of limiting factors was the assumption that the nutrients would be used for the growth and not for the accumulation of PHA otherwise. However, very recently, Grousseau *et al.* (2013), while studying the influence of phosphorus limitation on PHA production by *C. necator*, have shown that sustaining residual growth improves PHA production.[21]

Furthermore, the loss of nutrient-limiting conditions was also studied by some teams. Thus, Passanha *et al.* (2013) showed that different elements were vital for the initial growth and PHA accumulation,[22] whereas Berezina (2013) showed that *C. necator* is able to accumulate PHA up to 34% on reach medium and even without a specific carbon source for PHA accumulation.[23] Also, Bormann *et al.* (1998)[24] and Berezina *et al.* (2008)[25] showed that PHA content in cells during the accumulation step is not altered by the presence of nutrient-sufficient conditions in the reaction media.

In terms of PHA content in cells, the best results so far (90%) were reported by Passanha *et al.* (2013) when culturing *C. necator* on filtered digestate liquor,[22] whereas the best yield on a carbon source, 0.85 g g^{-1}, was attained by Cruz Pradella *et al.* (2012) on soybean oil.[26] Finally, the best productivity on a noble carbon source of 3.14 g L^{-1} h^{-1} obtained by Ryu *et al.* in 1997 remains the highest in the field[20] and was only approached by Shang *et al.* in 2003.[27] The best productivity on a waste carbon source, 2.5 g L^{-1} h^{-1}, was reported by Cruz Pradella *et al.* (2012) on soybean oil.[26]

It is, however, worth noticing that the extensive study of glycerol, mainly pure but also waste, is mainly due to the assumption that this chemical will flood the market in the very near future, being a by-product of bio-diesel production. However, until now, glycerol has been more expensive than glucose, and thus the economical considerations cannot justify the choice of this carbon source. Considering the sustainability approach, bio-fuels in general, and bio-diesel in particular, are no longer considered as promising ecological replacements of traditional fossil resources, as they bring up several issues, such as the occupation of arable lands, while the increasing exploration of shale gases, including the so-called "wet gazes" such as butane and propane can deeply change our perception of the availability of raw materials in the near future.

Other PHA$_{SCL}$ producing strains are *Pseudomonas fluorescens*, *Gamma proteobacterium* and *Haloferax mediterranei*, among others. Although they present interesting results in terms of carbon source used, such as sugarcane molasses or vinasse, the productivity, PHA content in cells and yield on carbon sources are often much lower than those reported for the *Cupriavidus* genus (Table 2.2).

Another strategy consists of using mixed cultures, such as activated sludge, to perform PHA accumulation. The important advantage of such a technique is the cheaper processes involved: no sterilization is required, the use of waste is generally applied and the infrastructure is almost always available at waste treatment facilities. Thus, even if the productivities are six-fold lower than those observed for the *Cupriavidus* genus on a pure carbon source, the economical feasibility is not that different.

As a matter of fact, we have to notice that even in these cases, the question of the limiting factor for triggering PHA accumulation inside the studied strains and systems was elaborated. Thus, Chen *et al.* (2011) observed that a smaller C/N ratio was more favorable for PHA accumulation in 72 hours in a culture of *Gamma proteobacterium*, whereas a higher C/N ratio was more favorable for PHA accumulation for longer hours of cultivation, up to 150.[28]

Concerning the activated sludge systems, Moralejo-Garate *et al.* (2013) showed that the presence of ammonia during the PHA accumulation step was not damaging for PHA production.[29]

Table 2.2 Sources of PHA$_{SCL}$ for other strains.

Strain	Substrate	Limiting nutrient	PHA content (%)	Yield (g g^{-1})	Productivity (g L^{-1} h^{-1})	Ref.
Pseudomonas fluorescens	Glutamate for growth sugarcane liquor for PHA accumulation	-/-	70	n.d.	0.23	122
Gamma proteobacterium	Volatile fatty acids	N[a]	45	0.45	-/-	35
Haloferax mditerranei	Vinasse	-/-	70	0.87	0.21	122
Activated sludge	Glycerol	N[a]	80	0.34	-/-	36
Activated sludge	Acetate + propionate	-/-	n.d.	>0.5	n.d.	122
Activated sludge	Paper mill	N	48	0.11	n.d.	122
Mixed cultures	Sugarcane molasses	-/-	56	0.67	0.56	122
Mixed cultures	Cheese whey	-/-	65	0.85	0.37	41

n.d.: not determined;
[a]The influence of the limiting factor was a specific object of the study and is discussed in the chapter text.

2.3 Properties of PHAs

2.3.1 Thermo-mechanical Properties

The physical and material properties of PHAs are greatly influenced by their monomer composition and chemical structure *i.e.* the length of the pendant groups which extend from the polymer backbone, the chemical nature of these pendant groups and the distance between the ester linkages in the polymer.[4,30]

As mentioned before, the PHA can be primarily divided according to the size of the pendant group (the number of carbon atoms in the monomer units) into a short chain length, medium chain length and long chain length (Figure 2.1).[31]

Besides its general classification (regarding the length of the side chain), there are several different types of side chain in the same family of PHA; they can also have functional groups attached. Of special interest are the 3-hydroxyl group and the carboxylic group of the functionalized groups in the side chain that allow further chemical modification, *e.g.*, halogens, carboxyl, hydroxyl, epoxy, phenoxy, cyanophenoxy, nitrophenoxy, thiophenoxy, and methylester groups.[9]

There are some strategies to target the final PHA structures to achieve tailor-made compositions. It can be relatively well controlled in three ways: (i) by choosing the correct microorganism host; (ii) by choosing appropriate cultivation conditions and carbon sources and/or (iii) by performing chemical modification after their production and purification. Combining these three strategies can bring success in achieving the desired composition.[32]

A comparison of the physical properties of some PHAs with polypropylene is given in Table 2.3.

2.3.2 Biocompatibility Properties

An important potential for PHA usage in the field of biocompatible materials has been foreseen for a long time. Their potential applications lie in wound management, the vascular system, orthopaedics, drug delivery, urology, *etc.*[33,34]

PHA_{MCL} are usually more biocompatible than PHA_{SCL}. Indeed, recent studies on PHA degradation products have shown that the cytotoxicity of the oligo-hydroxyalkanoates (OHA) decreased with the increasing OHA side chain length.[57] However, the most abundant studies have concerned P4HA and the PHA_{SCL}-*co*-PHA_{MCL} due to their larger accessibility and specific properties.

Studies of the PHB-*co*-PHHx copolymer have shown that a moderate content of 3-HHx (12%) co-monomers induces better adhesion of smooth muscle cells from rabbit aorta (RaSMCs), whereas a higher content of 3-HHx (20%) is found to be more suitable for the proliferation of these cells.[35]

Table 2.3 Physical properties of PHAs and some polymer commodities, adapted from. Hazer and Steinbüchel.[10,a]

Polyester/polymer	T_g (°C)	T_m (°C)	Crystallinity (%)	Elongation at break (%)	Young's modulus (MPa)	Ref.
PHB	7,15	175	50–80	5	1510	122
PHBV	− 15, −1.3	110, 112, 118, 157	56, 53	7	2000	122
PHB-*co*-20mol%3HV	− 1	145		50		10,15
PHB-*co*-10mol%3HHx	− 1	127				15
PHB-*co*-17mol%3HHx	− 2	120		850		15
PHB-*co*-47mol%3HV-*co*-16 mol%4HV-*co*-15 mol%3HHx-*co*-2 mol%3HO	− 15	118		1000		15,122
PHO-*co*-36mol%D-*co*-22.5mol%DO	− 44.5			40–70	9–15	21
P3MB	8	100				10
P4HB	− 40	53		1000		10
PHB-*co*-16mol%4HB			43	444		10
P3HPE	− 11	63				10
PHO	− 36	59,61				10
PHN	− 39, −29	48,54,58				10
P3H6MN	n.d.	65				10
PH-p5TV	17	95				10
PH6PHx	4	n.d.				10
PH5PoxV	14	n.d.				10
PH-*p*-nitroPV	28	n.d.				10
PH8-pMPoxO	14	97				10
PLA	45–60	150–162	2.5–6	3.6	1603	122
PP	− 15	176	50	400	1300	15,122
LDPE	− 75	104–113		200–750	160	122

[a]T_g is the glass transition temperature; T_m is the melting temperature; PHO-*co*-36mol%D = polyhydroxyoctanoate containing 36 mol% decanoate and 22.5 mol% dodecanoate; PLA = poly(lactic acid); PP = polypropylene; LDPE = low density polyethylene; n.d. = not determined.

Blended polyesters PHB/PHB-*co*-PHHx have shown strong proliferation of chondrocytes released from the cartilage slices of white rabbits from New Zealand.[36] Also, highly hydrophobic PHB-*co*-PHV-*co*-PHHx films were found to significantly increase the adhesion of human bone marrow mesenchymal stem cells (MSCs) compared to PLA and tissue culture plates (TCPs).[37] On the contrary, the hydrophobicity of the PHB-*co*-PHHx films was found to be responsible for the lack of human embryo lung fibroblast (HELF) cell attachment and proliferation.[38]

A comparison of the bioadsorption properties of electrospun PHB, PHB-*co*-PHHx and PHB-*co*-P4HB has shown that the proportion and type of the second monomer are of crucial importance in subcutaneous implantation in rats. Thus, the tissue response was significantly improved with the high content of 4-hydroxybutyrate co-monomer.[39] Also, the extremely high flexibility (1000%), low melting point and fairly good resistance to moisture of P4HB, even compared to polylactide (PLA) and polyglycolide (PGA), make this biomaterial a good candidate for heart valves, vascular grafts and suture textile products.[40,41]

Several techniques were used for the improvement of PHA-based biomaterials, some of them consisting of PHA structure modification, others mostly based on the surface treatment of obtained films.

The UV treatment of PHB-*co*-PHHx was found to induce a significant molecular weight (Mw) loss and broad distribution. This has contributed to a better biocompatibility, increasing the growth of mouse fibroblast cell lines L929.[42] Another study concerned the blending of PHB-*co*-PHHx with poly(propylene carbonate) (PPC).[43] The authors observed a significant improvement in mechanical properties (elastic modulus has passed from 93 to 50 MPa and the elongation to break increased from 16 to 340%). Also, wettability, the adsorption of fibronectin (FN) and the adhesion of RaSMCs have been improved by this treatment. The blending of PHBV with PLA or polyethylene glycol (PEG) has shown an improvement in the biocompatibility of PHB with FN and Chinese hamster lung (CHL) fibroblast cells respectively.[44,45]

The surface modification of PHAs was performed by either a direct treatment (plasma, NaOH, enzyme) or by coating. Among the direct surface treatments, the use of plasma was by far the most popular until now.[46–49] However, in any cases of plasma, lipase[46] or NaOH[50] treatment, the results were similar: enhancement of hydrophilicity with the concomitant improvement of biocompatibility. The only drawback was noticed in the case of NaOH treatment and concerned the aging effect, with the hydrophilicity decreasing with time.[52]

Different types of coating were also tested, with collagen and hyaluronic acid treatments being the most studied.[46,51] In this case, even if the hydrophilicity was generally improved, this was not automatically followed by an increase in biocompatibility.[53] Finally, the combination of plasma treatment and FN coating was found to give the most interesting results, especially in the proliferation of human umbilical vein endothelial cells (HUVECs).[51]

2.4   Structure of PHAs

2.4.1   PHA Structure Related to Microorganism Host and Carbon Substrate

The variability of bacterial PHAs that can be directly produced by fermentation is extraordinary large with more than 150 different hydroxyalkanoic acids, and even mercaptoalkanoic acid, that are known as constituents of these PHAs.[10] The production of PHA_{SCL}, PHA_{MCL} or PHA_{LCL} is directly linked to the microorganism host and the specificities of the enzymes (PHA synthases). PHAs can be synthesized as homopolymers or copolymers or even their blends, depending on bacterial strains employed or growth substrates, especially the PHA synthase that the strain contains.[52]

For instance, *Ralstonia eutropha* is well established as one of the best known bacteria among PHA_{SCL}-producing microorganisms and has several advantages for efficient PHA production as a recombinant host strain.[53,54] *R. eutropha* can intrinsically synthesize PHAs from various renewable inexpensive carbon sources, sugars and plant oils. However *R. eutropha* produces only PHA_{SCL}.

Depending on the subunit compositions and substrate specificities of the enzymes, PHA synthases are generally classified into four groups: class I, II, III, and IV[14]. Class I and II PHA synthases are composed of one subunit enzyme, PhaC. Class I PHA synthases such as *Ralstonia eutropha* and *Alcaligenes latus*, have molecular weights (Mw) between 61 and 68 kDa. According to their *in vivo* substrate specificity, type I PHA synthases preferentially utilize coenzyme A thioesters of various $3HA_{SCL}$ comprising three to five carbon atoms,[55] while class II PHA synthases mainly from pseudomonas display substrate specificity towards $3HA_{MCL}$. Some class II PHA synthases from *Pseudomonas sp.* 61–3 and *Pseudomonas sp.* 6–19 accept both SCL- and MCL monomers, with weak activity towards $_{SCL}$monomers. Class III PHA synthases are composed of two different subunits, PhaC and PhaE. These subunits have marginal sequence homology to class I and II PHA synthases; the PhaC subunit displays only 20–30% homology, for example. Class III PHA synthases are highly specific for $3HA_{SCL}$, but do accept $3HA_{MCL}$ when expressed in some pseudomonas. Class IV PHA synthases are composed of two different subunits, PhaC and PhaR, which are usually found in Bacillus strains producing PHB. Various HA-CoAs are the substrates of the PHA synthases and undergo polymerization to PHA polymer granules in the cells.[14,56–58] Table 2.4 shows representative PHA synthases used for the synthesis of PHA consisting of both SCL and MCL monomers.[14]

In spite of PHB being considered an environmental friendly polymer with similar material properties to polypropylene (PP), it has not been used on a large scale to replace conventional polymers because it presents some drawbacks in its mechanical properties. Considering polymer mechanical properties, it is important to consider three basic properties when comparing the usefulness of a polymer for a given commodity application. The elongation at failure is a measurement of toughness and reflects the total deformation that the polymer can withstand before fracture. The Young's modulus is a measure of stiffness that considers the slope of the stress response to deformation at very low strains. Steeper slopes at low deformations are indicative of stiffer materials. Finally, the ultimate tensile strength is a measure of the maximum strength of the material prior to the onset of

Table 2.4 Representative PHA synthases used for the synthesis of PHAs consisting of both SCL and MCL monomers.[14]

Strain	PHA synthase	Characteristics	Monomer composition
Pseudomonas sp. 61-3	*phaC1*	Natural	3HB, 3HHx, 3HO, 3HD, 3HDD
	phaC1	Combinatorial mutations in E130, S325, S477, Q481, Q508	LA, 3HB, 3HHx, 3HO, 3HD, 3HDD
Pseudomonas sp. MBEL 6-19	*phaC2*	Natural	3HB, 3HHx, 3HO, 3HD, 3HDD
	phaC1	Combinatorial mutations in E130, S325, S477, Q481	LA, 3HB, 3HHx, 3HO, 3HD, 3HDD
P. aeruginosa	*phaC1*	Combinatorial mutations in E130, S325, S477, Q481,	LA, 3HB, 3HHx, 3HO, 3HD, 3HDD
P. putida	*phaC1*	Combinatorial mutations in E130, S325, S477, Q481,	LA, 3HB, 3HHx, 3HO, 3HD, 3HDD
P. chlororaphis	*phaC1*	Combinatorial mutations in E130, S325, S477, Q481	LA, 3HB, 3HHx, 3HO, 3HD, 3HDD
P. resinovorans	*phaC1*	Combinatorial mutations in E130, S325, S477, Q481	LA, 3HB, 3HHx, 3HO, 3HD, 3HDD
A. caviae	*phaC*	Natural	3HB, 3HHx
	phaC	Combinatorial mutations in N149, D171, A505	3HB, 3HHx, 3HO
A. hydrophila	*phaC*	Natural	3HB, 3HHx
Rhodococcus aetherivorans	*phaC*	Natural	3HB, 3HHx

plastic deformation. This value is obtained from the maximum of the initial peak in the stress *versus* strain diagram. While the modulus (stiffness) and ultimate tensile strength (overall strength) of PHB and related copolymers are suitable for many commodity-type uses, a major setback in the commercialization of these materials relates to the extremely brittle character of these polymers and very low flexibility.[59,60] For instance, PHB and PHBV materials have been shown to fracture at less than 10% elongation.[18]

Also, it is hard to process PHB due to its high melting temperature of approximately 170 °C, which is very near to its degradation temperature. Therefore, a solution to these drawbacks could be the copolymerization of 3-HB with other monomers that confer less stiffness and tougher properties (which bestow greater flexibility and lessen breakage) and to reduce the melting point. PHA copolymers such as poly(hydroxybutyrate-*co*-3-hydroxyvalerate) [P(3HB-*co*-3HV)], poly(hydroxybutyrate-*co*-4-hydroxybutyrate) [P(3HB-co-4HB)], poly(hydroxybutyrate-*co*-3-hydroxyhexanoate) [P(3HB-*co*-3HHx)], and poly(hydroxybutyrate-*co*-3-hydroxyalkanoate) [P(3HB-*co*-3HA)] have been examined extensively as candidates with such material properties.[14,61] Because class I, III, and IV PHA synthases have the ability to naturally accept 3HBCoA, 3HV-CoA, and 4HB-CoA as substrates, incorporation of these sclmonomers into the PHB polymer backbone has been carried out by employing natural PHA synthases from *R. eutropha*, *A. latus*, and *Allochromatium vinosum*. However, the incorporation of MCL monomers into the P3HB backbone is not efficient, since substrate specificities of natural PHA synthases are highly specific in terms of carbon chain length of HA-CoAs and accept only the SCL or MCL monomer. Therefore, it was considered necessary to screen natural PHA synthases or to evolve PHA synthases to accept both SCL and MCL monomers for the production of PHB copolymers containing a small amount of MCL monomers.[14]

4-Hydroxybutyrate (4-HB) was produced by *Aeromonas hydrophila* 4AK4, *Escherichia coli* S17-1, or *Pseudomonas putida* KT2442 harboring 1,3-propanediol dehydrogenase gene dhaT and aldehyde dehydrogenase gene *aldD* from *P. putida* KT2442, which are capable of transforming 1,4-butanediol (1,4-BD) to 4HB. 4HB containing fermentation broth was used to produce homopolymer poly-4-hydroxybutyrate [P(4HB)] and copolymers poly(3-hydroxybutyrate-co-4-hydroxybutyrate) [P(3HB-4HB)]. Recombinant *A. hydrophila* 4AK4 harboring plasmid *pZL-dhaT-aldD* containing *dhaT* and *aldD* was the most effective 4HB producer, achieving approximately 4 g l^{-1} 4HB from 10 g l^{-1} 1,4-BD after 48 h of incubation. The strain produced over 10 g l^{-1} 4HB from 20 g l^{-1} 1,4-BD after 52 h of cultivation in a 6-L fermenter. Recombinant *E. coli* S17-1 grown on 4HB containing fermentation broth was found to accumulate 83 wt% of intracellular P(4HB) in a shake flask study. Recombinant *Ralstonia eutropha* H16 grew to over 6 g l^{-1} cell dry weight containing 49 wt% P(3HB-13%4HB) after 72 h.[8]

PHA accumulating bacteria can also synthesize copolymers containing 3-mercaptopropionate (3MP), 3-mercaptobutyrate (3MB), 3-mercaptovalerate (3MV), in addition to 3HB, representing the first examples of polythioesters (PTE) as a novel class of biopolymers. PMP, PMB, and PMV homopolymers were produced employing a recombinant strain of *Escherichia coli* expressing a non-natural PTE biosynthesis pathway.[62] The PTE structure is shown in Figure 2.3 and the T_m values are 170 °C for PMP, 100 °C for PMB, and 84 °C for PMV, whereas the T_m values are 77 °C for PHP, 175 °C for PHB, and 112 °C for PHV, respectively.[10]

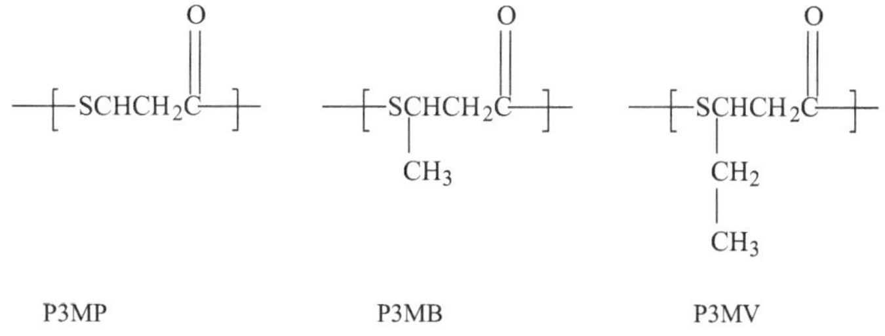

Figure 2.3 Chemical structure of poly(3-mercaptoalkanoate)s: P3MP, P3MB and P3MV.

Besides the microorganisms, the carbon source used in PHA production is connected with the functionalities present in the side groups. Functionalized PHAs are obtained either by feeding structurally related substrates processed through the beta-oxidation pathway, or using specific strains able to transform sugars or glycerol into unsaturated PHAs by *de novo* fatty-acid biosynthesis. Functionalized PHA_{MCL} provide modified mechanical and thermal properties, and consequently have new processing requirements and highly diverse potential applications in emergent fields such as bio-medicine.[63] In this respect, attention has been given to producing PHAs bearing terminal double bonds in their side chains. The reason behind the preference for unsaturated PHAs is due to the possibility of establishing high-yield production processes.[64] Alkylic substrates containing double bonds are cheaper and generally exhibit less toxicity compared to substrates with reactive functional groups. At the same time, unsaturated side chains of PHA are susceptible to chemical modifications. Starting from alkene func-tion, PHAs have been modified to exhibit functionalities *en route* towards new PHA-based biomaterials.[42,65]

One example of unsaturated PHA_{SCL} is poly(3-hydroxy-4-pentenoic acid), PHPE. *Burkholderia sp.* grown on a sucrose-containing mineral salts medium with phosphate limitation induces poly(hydroxyalkanoate) (PHA) accumu-lation. Under these conditions, the cultures accumulated 3-hydroxybutyric acid (3HB) and 3-hydroxy-4-pentenoic acid (3HPE) containing polyesters. Solvent fractionation of the purified polyester indicated the presence of two homopolymers, poly(3HB) and poly(3HPE), rather than a co-polyester with random monomer distribution as has been reported previously.[66] The PHPE homopolyester compared to PHB and PHV has very low crystallinity and a lower melting temperature: $T_g = -10.8$ °C and $T_m = 63$ °C.[44] Incorporation of 3HPE in a P<HB-*co*-HV-*co*-HPE) terpolyester can be achieved when *Rho-dospirillum rubrum* is grown on 4-pentenoic acid and on an equimolar mixture of 4-pentenoic acid and *n*-pentanoic acid.[67,68]

PHAs containing double bonds can also be produced by recombinant *Methylobacterium extorquens* strains when fed unsaturated fatty acids.[69] These PHA comprise PHA_{SCL} and PHA_{MCL}. One reason for choosing a methylotrophic microorganism for such a purpose was that an important portion of the production process would use methanol, a simple, inexpen-sive, very abundant, and non-food substrate.[47] Another unsaturated poly-ester such as poly(3-hydroxy-10-undecenoate), PHU, and others, are produced by various pseudomonas when cultivation is done containing as substrates animal or vegetable edible oils, oily acids, 10-undecenois acid and 1-alkenes.[70–72] The alkyl side-chains of the produced PHA contained varying degrees of unsaturation. PHA from coconut oil was composed entirely of saturated side-chains, whereas PHA-soy contained 4.2 mol% olefinic groups in its side-chains.[73]

Pseudomonas oleovorans can be grown on *n*-alkanoates and related carbon sources such as alkanes to produce PHA_{MCL} that are all copolymers.[74] *P. oleovorans* was grown in homogeneous media containing *n*-alkanoic acids, from formate to decanoate, as the sole carbon sources. Formation of intracellular poly(β-hydroxyalkanoates) was observed for hexanoate and the higher *n*-alkanoic acids. In most cases, the major repeating unit in the polymer had the same chain length as the *n*-alkanoic acid used for growth, but units with two carbon atoms fewer or more than the acid used as a carbon source were also generally present in the polyesters formed. Co-polymers containing as many as six different types of β-hydroxyalkanoate units were formed.[75] *P. oleovorans* was also used to produce methyl-branched PHA such as poly(3-hydroxy-6-methylnonanoate), P(H6MN), when grown on mixtures of methyloctanoates with *n*-octanoate. The polyesters obtained from 7-methyloctanoate and from its mixtures with *n*-octanoate contained units with the methyl branches in the pendant group, as did the copolymers from mixtures of 5- and 6-methyloctanoate with *n*-octanoate. The average molecular weights of the copolyesters produced were in the range of 220 000 to 410 000, with Mw/Mn ratios of 1.7 to 1.9.[76]

Choi and Yoon (1994) tested forty-two different carbon sources for poly-ester synthesis by *Pseudomonas citronellolis* (ATCC 13674). These included linear C_2 to C_{10} monocarboxylic acids, C_3 to C_{10} dicarboxylic acids, saccharides, α,ϖ-diols, hydrocarbons, and 3-methyl-branched substrates such as 3,7-dimethyl-6-octen-1-ol (citronellol), 3-methyl-*n*-valerate, 3-methyl-1butanol, and 3-methyladipate.[77] Polyesters from nine monocarboxylic acids and two related carbon sources could be metabolically divided into three groups. The first group of C_2 to C_4 carbon sources resulted in copolyesters composed of 61 to 70 mol% 3-hydroxydecanoate, 23 to 33 mol% 3-hydroxyoctanoate, 3.6 to 9.0 mol% 3-hydroxy-5-*cis*-dodecenoate, and 1.8 to 2.6 mol% 3-hydroxy-7-*cis*-tetradecenoate. Carbon sources in group II (C_7 to C_{10}) produced copolyesters composed of 3-hydroxyacid monomer units with the same number of carbon atoms as the substrate (major constituent) and monomer units with either two fewer or two more carbons. Negligible amounts of 3-hydroxy-5-*cis*-dodecenoate and 3-hydroxy-7-*cis*-tetradecenoate were detected in copolyesters from this group. Copolyesters from group III (C_5 and C_6) had a monomer unit distribution that could be said to be between those of groups I and II. In addition, a novel copolyester, poly(3-hydroxy-7-methyl-6-octenoate-*co*-3-hydroxy-5-methylhexanoate), was synthesized when grown on citronellol.[55]

A number of *Pseudomonas* strains can accumulate PHAs from a variety of aromatic hydrocarbons. In many strains, the level of PHA accumulation is dependent on the side chain length of the phenylalkanoic acid provided for growth.[78] The PHA accumulated from styrene and phenylacetic acid was composed of aliphatic monomers only. The PHA accumulated from any one of the phenylalkanoic acids with five carbons or more in their side chain was almost identical for all strains with the PHA composed of both aromatic and aliphatic monomers. The predominant monomers accumulated were 3-hydroxyphenylvaleric acid and 3-hydroxyphenylhexanoic acid. The addition of the metabolic pathway inhibitors acrylic acid and 2-bromooctanoic acid resulted in decreased levels of PHA from phenylacetic acid, suggesting a role for both b-oxidation and fatty acid synthesis in PHA accumulation from phenylacetic acid.[56] PHAs bearing phenoxy groups in the side chains have been produced by *Pseudomonas oleovorans.* Phenoxyundecanoic acid was used as a single carbon source for the bacterial culture and the polyester obtained was a copolymer: poly(3-hydroxy-5-phenoxypentanoate-*co*-3-hydroxy-9-phenoxynonanoate) (5:6:1 mole ratio).[79] Other carbon substrates (ω-phenoxyalkanoates) can also be used.[80]

PHAs containing repeating units with terminal epoxide groups were obtained when C7–C12 1-alkenes were fed separately as the only carbon source for *Pseudomonas cichorii* YN2. The content of epoxidized units in the polymers was in the range of 4–20 mol%, which was not dependent on the C atom length of the 1-alkene used as a substrate. The polymers produced undergo a glass transition at around −40 °C, and number average molecular weights were in the range of 150 000–200 000 as determined by GPC relative to polystyrene, with Mw/Mn ratios of 1.9 to 2.5. As an intermediate, the corresponding 1,2-epoxyalkane was found in the culture medium. According to this result, the epoxidation of the 1-alkene is the initial step in the synthetic pathway of the epoxy unit in the polymer (Figure 2.4).[81]

Polymers containing nitrophenyl groups were isolated when *P. oleovorans* was grown with 5-(2′,4′-dinitrophenyl)valeric acid (DNPVA) containing variable ratios of this and nonanoic acid (NA). The bacteria produced polymeric materials with 1.2 – 6.9% of repeating units containing 4′-nitro and/or 2′,4′-nitrophenyl rings, depending on the DNPVA : NA molar ratio. The polymeric material isolated from approximately 15% dry cell weight when using a 7:3 DNPVA : NA molar ratio at 10 mM total concentration was a yellow, elastic substance. Thermal analysis indicated the presence of two T_g values, T_g,1 = − 35.95 °C and T_g,2 = 28.74 °C, and one T_m value of 56.42 °C. These data suggest the presence of two polymers, one which contains nitrophenyl rings and the other being the copolymer of nonanoic acid.[82]

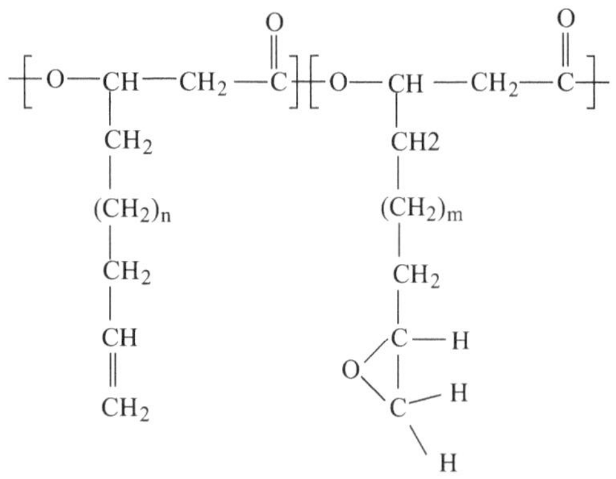

Figure 2.4 PHA bearing an epoxide group.

2.4.2 Chemical Modification of PHA Structure

Some PHA_{SCL} may be too rigid and brittle and may lack the superior mechanical properties required for biomedical and packaging film applications. In contrast, as described before, PHA_{MCL} may be elastomeric but have very low mechanical strength. It is reported that the stress–strain curve for P(3-hydroxy-octanoate) containing minor quantities of other MCL monomers was typical of those observed for elastomers with no yield stress appearing in the stress–strain curve. A Young's modulus of 17 MPa and an elongation to break of 250–350% were also determined.[83]

Therefore, for packaging materials, biomedical applications, tissue engineering, and other specific applications, the physical and mechanical properties of microbial polyesters need to be diversified and improved.[5,10] The mechanical properties of PHA are directly correlated with its structure and crystallinity. An increase in the variety of side chains within one polymer chain of PHA_{MCL} can modify its ability to crystallize and as a consequence there are some distinct differences in the crystallinity of PHA_{MCL}.[4] Obtaining a low crystallinity could occur once the polymers have attached large and irregular pendant side groups. These groups inhibit the close packing of the polymeric chains in a regular three-dimensional fashion to form a crystalline array.

Once the axial geometry in a polymeric chain is the main factor in determining the ability of a chain to form crystallites, the crystalline contribution is probably due to isotactic and syndiotactic structure sequences.[4,84,85] In fact, saturated PHA_{MCL}, which are able to crystallise due to their isotactic configuration, are also seen to crystallise with alkyl side chains in an extended conformation to form ordered sheets, but they still show a reduced degree of crystallinity when compared to P(3HB) or P(3HB-co-3HV), owing to low crystallization rates.[4]

The chemical modification of PHA can be done *via* block copolymerization and grafting reactions, chlorination, cross-linking, epoxidation, hydroxyl and carboxylic acid functionalization and so on. With the help of chemical modification it is possible to prepare a wide range of polymers, from hydrophobic to hydrophilic with high stability by adjusting the nature, structure and percentage of grafted side groups.[86]

A common route that can be applied for obtaining polyhydroxyalkanoates with desirable functionalities is to produce PHAs with terminal double bonds followed by chemical modification steps. Carbon double bonds are comparatively inert but can be easily transformed into reactive functional groups under mild reaction conditions. Non-functionalized PHAs can also be activated by surface modification techniques. The resulting tailor-made structural and material properties have positioned polyhydroxyalkanoates well to contribute to the manufacturing of second and third generation biomaterials.[43]

Unsaturated PHA_{MCL}, such as polyhydroxyundecenoate (PHU), have very low crystallinity and cross link when irradiated with UV light, even in the absence of photosensitizers or photoinitiators. Cross-linking of these PHA$_{MCL}$ is expected to yield useful biomaterials, such as biodegradable rubbers. In addition, since cross-linked PHAs are less susceptible to PHA depolymerase than natural PHAs, these polymers can be useful as photosensitive materials for microlithography that are also environmentally friendly.[87]

Graft copolymerization is a technique used to transform the properties of natural polymers. Block or graft copolymerization is the insertion of different polymer segments into an existing polymer backbone or at the side chain of an existing polymer, respectively.

Due to their biocompatibility, PHAs proved to be good candidates for biomedical applications including biomedical devices, biodegradable drug carriers or tissue engineering. However, the intrinsic hydrophobic properties of PHAs restrict their applications as cell growth supports. So, a graft copolymer of PHA and a marine exopolysaccharide (EPS), HE800, were synthesized in order to improve the compatibility between hydrophobic PHBHV and hydrophilic HE800. In the grafting method, the carboxylic functions of PHBHV were activated with acyl chloride functions, allowing coupling to the hydroxyl groups of HE800 (Figure 2.5).[88]

Dextran-*graft*-PHBHV copolymers were prepared using click chemistry. Well defined and functional dextran backbones containing azide groups had been prepared by tosylation and a subsequent nucleophilic displacement

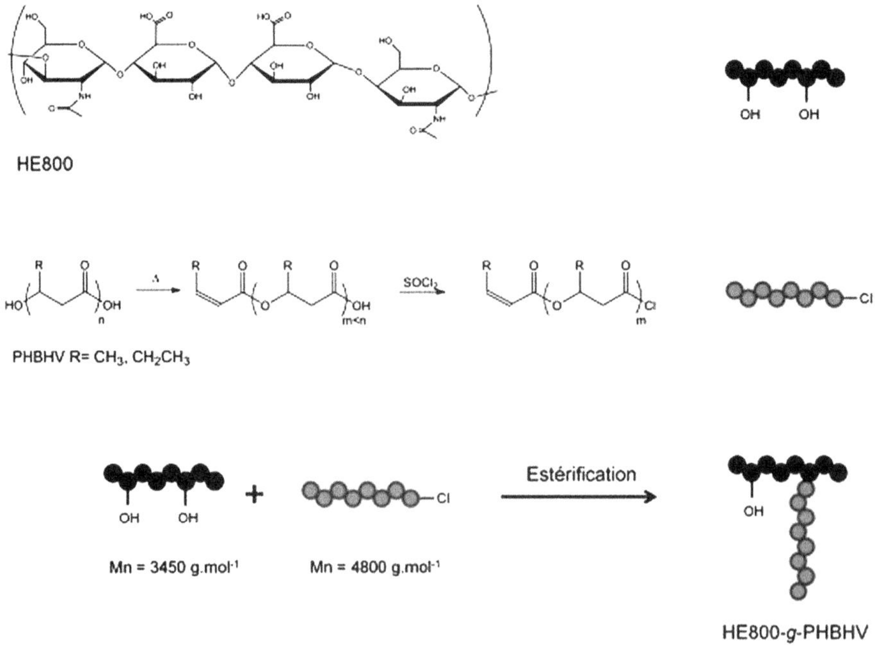

Figure 2.5   Preparation of the copolymer HE800-*g*-PHBHV.

reaction with sodium azide (DSN3 = 1). Well defined PHBHV oligomers containing an alkyne end group were prepared in a one-step reaction by direct alcoholysis from natural polyesters using propargyl alcohol with dibutyltin dilaurate as a catalyst. The presence of PHBHV on the dextran backbone led to the formation of stable nanoparticles (160 nm) without surfactant by an emulsion–solvent evaporation method.[89]

Poly(ethylene oxide) (PEO) as a hydrophilic and biocompatible polyether is widely used in biomedical applications.[90] An amphiphilic triblock copolymer, an enantiomerically pure telechelic OH-terminated poly[(R)-3-hydroxyoctanoate] (PHO-diol), poly[(R)-3-hydroxyoctanoate]-*co*-poly[(R)-3-hydroxy-7-oxooctanoate] (PHOO-diol), and poly[(R)-3-hydro-xyoctanoate]-*co*-poly[(R)-3-hydroxy-7-octenoate] (PHUO-diol) have been synthesized by catalytic transesterification with ethylene glycol.[91] The number average molecular weights (Mn) of these telechelic diols reached $(2.0 - 3.0) \times 10^3$. For PHOO-diol and PHUO-diol, the side chain functional groups remained, which provides additional reactive groups for further polymerization or modification. The glass transition temperatures (T_g) of the telechelic diols are between −46 and −56 °C and the melting transition temperatures (T_m) are lower than 40 °C, all determined by DSC.[92,93]

Aliphatic polyesters coupled with monomethoxy poly(ethyleneoxide), mPEO, as PHB-*b*-PEO diblock copolymers are often used as drug delivery systems, and many applications in this field involve polylactides (PLA) coupled to mPEO because of the biodegradability and biocompatibility of PLA. Three and four-arm star polyethylene oxide-polylactide copolymers (*s*-PEO-PLA) can be synthesized by the use of triethanolamine and pentaer-ythritol and initiating agent, respectively.[94] Similarly, because PHB is also a chiral aliphatic polyester, PHB-bmPEO diblock copolymer can be synthesized. For this, PHB and poly(ethylene glycol) methyl ether are melted under vacuum at 190 °C in the presence of bis(2-ethylhexanoate) tin catalyst.[95]

New diblock copolymers of selected PHAs (PHB, PHBV, PHO) with atactic poly[(R,S)-3-hydroxybutyrate] (a-PHB) were obtained by using low-molecular weight macroinitiators derived from natural poly(3-hydro-xyalkanoates), which contain olefinic units and activated by 18-crown-6 ether carboxylic end groups. These new polymers are suitable for use as blend compatibilizers.[96]

A chemical synthesis of functionalized atactic poly(3-hydroxybutanoic acid) and its copolymers with the aid of activated anionic initiators as well as enzymatic synthesis of PHB copolymers (using lipase PPL) are described by Jedliński *et al.* (1999). Using these new synthetic approaches, PHBs with defined chemical structures of the end groups as well as block, graft and random copolymers have been obtained.[97] The chemical synthesis of co-polymers containing structural segments derived from natural origin PHAs is based on the concept of partial depolymerization of a natural polyester to oligomers containing end groups suitable for initiation of polymerization of other monomers.[75,98]

The terminal carboxyl groups of PHAs are open to react with chitosan amine functions and cellulose hydroxyl functions.[99] For the synthesis of PHB-*g*-chitosan graft copolymers, chitosan solutions in dilute acetic acid are treated with different molar ratios of reduced-molecular-weight PHB. The partially polymerized PHB samples can be prepared either *in situ* or before use by dissolving the PHB in a mixture of acetic acid–DMSO (1 : 50, v/v) and stirring for 16 h at ambient temperature.[77] Although neither of the parent polymers is water soluble, the PHA-chitosan derivatives form opaque, viscous solutions in water.[10,77] Upon drying of such solutions, strong elastic films can be prepared. The T_m of PHB shifts from 175 to about 150 °C for PHB-*g*-chitosan. At the same time, the endotherm of chitosan also decreased from 116 to 105 °C.

PHO, poly(3-hydroxybutyrate-*co*-3-hydroxyvalerate) (PHBV), and linoleic acid can be grafted onto chitosan *via* condensation reactions between carboxylic acids and amine groups.[78] The percentage of microbial polyester grafted onto the chitosan backbone varied from 7 to 52 wt% as a function of the molecular weight of PHA, namely as a function of steric effect. The plasticizer effect of PHO in PHO-*g*-chitosan lowered the T_m of the graft copolymer to 80 °C depending on its PHO content. Thermal analysis of PHO-g-chitosan graft copolymers indicated the plasticizer effect of PHO by showing melting transitions, T_ms, at 80, 100, and 113 °C or a broad T_ms between 60.5–124.5 °C and 75–125 °C while pure chitosan showed a sharp T_m at 123 °C.[100]

Poly(ester urethane)s with P3HB as the hard and hydrophobic segment and poly(ethylene glycol) (PEG) as the soft and hydrophilic segment were synthesized from telechelic hydroxylated PHB (PHB-diol) and PEG using 1,6-hexamethylene diisocyanate as a non-toxic coupling reagent.[101] The PHB segment and PEG segment in the poly(ester urethane)s formed separate crystalline phases with lower crystallinity and a lower melting point than those of their corresponding precursors, except that no PHB crystalline phase was observed in those with a relatively low PHB fraction. Thermogravimetric analysis showed that the poly(ester urethane)s had better thermal stability than their precursors. Water contact angle measurements and water swelling analysis revealed that both the surface hydrophilicity and bulk hydrophilicity of the poly(ester urethane)s were enhanced by incorporating the PEG segment into PHB polymer chains. Regarding mechanical properties, it was found that the poly(ester urethane)s were ductile, while natural source PHB is brittle. The Young's modulus and the stress at break increased with increasing PHB segment length or PEG segment length, whereas the strain at break increased with increasing PEG segment length or decreasing PHB segment length.[101]

Hao and Deng (2001) prepared semi-interpenetrating networks (IPNs) hydrogels based on bacterial P3HB and net-poly(ethylene glycol) (net-PEG) by the UV irradiation technique. Net-PEG-based hydrogels all show higher equilibrium water contents (EWCs), the crystallinity of PEG segments is noticeably decreased by cross-linking and would drop further with increasing the amount of P3HB. Incorporation of a semi-IPN structure with PHB could significantly improve the mechanical properties of hydrogels when compared with that of pure net-PEG. The PHB/net-PEG (50 : 50) semi-IPN2, ranked as that possessing the best mechanical properties in the wet state, was expected to be useful as a biomedical material.[102] IPNs have tensile strength values varying from 2.5–8.5 MPa and have elongation at break values varying from 3.8 to 35.5%.

In another example, monoacrylate-poly(ethylene glycol) (PEGMA) monomer was grafted onto PHO by UV irradiation in a chloroform solution containing benzoyl peroxide to obtain PEGMA-grafted PHO (PEGMA-*g*-PHO) copolymers. The UV irradiation treatment is liable to cause scission of the chemical bonds and the generation of free radicals in both the PHO and the PEGMA.[103] The proteins and platelets had a significantly lower tendency to adhere to the PEGMA-*g*-PHO copolymers than to PLLA. The graft copolymer with a high grafting degree (DG) of PEGMA was very effective in reducing the protein adsorption and platelet adhesion and did not activate the platelets. In respect to the thermal properties of the PHO and PEGMA-*g*-PHO copolymers, PHO had the highest glass transition temperature (T_g), melting temperature (T_m), and enthalpy of fusion (ΔH_m), while the T_g, T_m, and ΔH_m of the graft copolymers decreased as the DG of the PEGMA groups in the copolymers increased. The results obtained in this study suggest that PEGMA-*g*-PHO copolymers have the potential to be used as blood-contacting devices in a broad range of biomedical applications.[81]

Chlorination of microbial polyesters P3HB and PHO was carried out by passing chlorine gas through their solutions. Usually, two different chemical reactions have been carried out: (1) an addition reaction when unsaturated PHAs are the polymers to be modified or (2) a substitution reaction if the polymers to be functionalized are saturated hydrocarbons.[104] In both cases, the hydrolysis of the PHA during the chlorination process is unavoidable. The chlorine contents in chlorinated P3HB (PHB-Cl) and chlorinated PHO (PHO-Cl) were between 5.45 and 23.81 wt% and 28.09 and 39.09 wt%, respectively. The thermal properties of PHO-Cl were dramatically changed with an increase in its glass transition ($T_g = 2$ °C) and the melting transition (T_m). The T_g of PHB-Cl varied from −20 to 10 °C, and its T_m decreased to 148 °C. The chlorinated poly(3-hydroxyalkanoate)s (PHA-Cl) were converted to their corresponding quaternary ammonium salts (PHA-N + R3), sodium sulfate salts (PHA-S), and phenyl derivatives (PHA-Ph). Cross-linked polymers were also formed by a Friedel − Crafts reaction between benzene and PHA-Cl.[105] In another study, unsaturated polyesters, (PHA − DB), obtained from pure anchovy (hamci), hazelnut, soybean oily acids and mixtures of octanoic acid (in weight ratios of 50 : 50 and 70 : 30) were chlorinated up to 54 wt% Cl content. The molecular weights (Mw) of the chlorinated PHAs were between 1.3×10^4 and 3.0×10^4 and decreased with the increase in chlorine content in PHA − DB. Melting transitions of the chlorinated PHA were between 62 and 125 °C depending on the chlorine content when compared with those of the original PHA, 44 − 55 °C.[82]

An enhancement in the elastic response of PHO, and poly(β-hydroxyoctanoic-*co*-undecylenic acid) (PHOU), was achieved using peroxide crosslinking both with and without multifunctional co-agents and curing thermally under vacuum to obtain cross-linked PHOU. Differential scanning calorimetry (DSC) showed that crosslinking could eliminate all crystallinity. The elastic response was improved and in general, the crosslinked materials exhibited a decrease in tensile modulus, and a very low tensile strength and tear resistance.[106] In another work, the enhancement in mechanical properties of PHOUs containing varying amounts of unsaturation was attempted through sulfur vulcanization. DSC showed that crosslinking could eliminate all crystallinity. The elastic response was near ideal with less than a 5% tensile set after 200% elongation. In general, sulfur-vulcanized materials exhibited a decrease in tensile modulus, tensile strength and tear resistance, which did not appear to vary with crosslink density. Unlike the case of peroxide crosslinking, material integrity was left intact and mechanical property results are reported. The network structure was elucidated through the determination of the molecular weight between crosslinks.[107] PHOU was crosslinked using gamma-irradiation. The advantage of this process is that it would allow sterilization and crosslinking simultaneously.[108]

Epoxidation of different bacterial polyesters containing unsaturated side chains in the repeating units with *m*-chloroperbenzoic acid, as a chemical reagent, led to quantitative conversions of the unsaturated groups into epoxy groups with no side reactions observed on the macromolecular chain by molecular weight measurements. It has been possible to produce new functional bacterial polyesters containing terminal epoxy groups in the side chains, in variable proportions up to 37%.[109] Epoxidation of the unsaturated side chains in PHOU, by reacting the polymer with *m*-chloroperbenzoic acid (MCPBA) in homogeneous solution, was readily carried out. The acidic conditions used for the epoxidation reaction did not result in a significant decrease in molecular weight of the PHOU. The epoxidized PHOUs were completely soluble, indicating that cross-linking did not occur, but transformation of the vinyl groups in PHOU into epoxide groups caused a decrease in the melting temperature and the enthalpy of melting. In contrast, the glass transition temperatures increased in a linear manner with epoxide group content in the product polymer, and the thermal degradation behavior of the epoxidized polymer was considerably different from that of the initial polymer.[110]

The synthesis of PHAs containing pendant diol groups was carried out by the chemical modification of unsaturated PHA using KMnO$_4$ in cold alkaline solution (pH 8–9) at 20 °C without a severe reduction in molecular weight.[111] The degree of hydroxylation increased to approximately 60% after 3 h of reaction, but there was no further increase for the longer reaction times, and the degree of hydroxylation of products was almost constant at 50–60% after 3 h, irrespective of the unsaturated unit content of the original PHOU or of the KMnO$_4$/unsaturated unit molar ratios, which varied from 0.7 to 2.0. The polymers, which were 40–60% hydroxylated, were completely soluble in polar solvents including an 80 : 20 acetone–water mixture, methanol and DMSO, indicating a considerably enhanced hydrophilicity of the modified PHA.[89]. Despite the fact that the $KMnO_4$ treatment allowed the hydroxylation of about half of the pendant vinyl groups present in the PHA, the reaction performed with 9-borobicyclononane (a hydroboration–oxidation reaction) hydroxylated most of the vinyl groups (almost 100%) present in poly[(R)-3-undecenoate]. Molecular weight and polydispersity of hydroxylated PHU were found to be 10 000 and 1.23, respectively, while those of the originals were 32 000 and 2.42. The decomposition temperatures of the original and hydroxylated PHU were 280 and 200 °C, respectively.[112]

The carboxylation of unsaturated PHA has been performed through reactions that involved the conversion of the double bonds to thioethers *via* the free-radical addition of 11-mercaptoundecanoic acid[113] or (R)-3-mercaptopropionic acid. Transesterification reactions of poly(3-hydroxy butyrate) were carried out under reflux of 1,2-dichlorobenzene in the presence of 1,4-butane diol, poly(ethylene glycol) bis(2-aminopropyl ether) with molecular weights of 1000 and 2000, poly(ethylene glycol)methacrylate or glycerol at 180 °C. Addition reactions of bromine and the –SH groups of 3-mercaptopropionic acid to the double bond of poly(3-hydroxy-10-undecenoate) were also carried out. The molecular weights of the modified polymers (despite the addition of mercapto acids to the double bonds) remained almost constant.[114]

Copolymers of PHAs containing repeating units with unsaturated or brominated pendant side chains have been obtained from cultures of *Pseudomonas oleovorans* grown on mixtures of octanoic acid and undecenoic acid or 11-bromoundecanoic acid as carbon sources. These polymers, bearing reactive functionalities, have been used to graft acetylated maltosyl units either by anti-Markovnikov addition to the double bond or SN2 substitution of the halogen. De-O-acetylation of the sugar moieties yielded PHAs with new properties.[115]

All the chemical modifications will significantly change the properties of PHAs. Table 2.5 shows some typical changes in properties resulting from the modification of PHA_{MCL}.

2.5 Outlook

Different metabolic pathways guide the production of different types of PHAs. Thus the choice of carbon source for the production of these PHAs has unequal importance for PHA_{SCL} and PHA_{MCL}. For PHA_{MCL}, the choice of the carbon source is crucial for the final structure of the biopolymer. On the contrary, for PHA_{SCL}, almost any carbon source can be used. Thus, more in depth studies in this field were performed for PHA_{SCL}. It was thus shown that the *Cupriviadus* genus has great potential for this type of production. Recent work on mixed microbial cultures has also shown an interesting perspective, especially knowing that these systems are usually cheaper to establish. Also, for a long time, the assumed hypothesis of the necessity of

Table 2.5 Typical changes in properties resulting from the modification of PHA$_{MCL}$.[65]

Modification methods	Product	Molecular weights		Thermal properties		Mechanical properties	
		$M_n \times 10^{-4}$	M_w/M_n	T_g(°C)	T_m(°C)	Tensile strength (MPa)	Elongation at break (%)
Blending	PHO/PLA50 (80/20, w/w)	8.0/0.1		−39.0/−10.0	49.0		
	Semi-IPN PHU/PLGA (80/20, w/w)	5.0/5.0				1.1	205.0
Crosslinking	Radiation treated PHA-tal	4.4~7.3	3.6~4.9	−41.7~−42.3	43.1~43.3	4.0~5.1	195.0~235.0
	Gamma-ray treated PHOU	1.1~8.2		−35.0~−40.0	49.0~59.0		
Chemical reaction	Chlorinated PHAs	1.0~2.0	1.3~1.7	2.0~58.0	104.0~134.0		
Grafting	Epoxidized PHAs	9.3~18.4	1.4~2.5	−28.0~−34.0	41~54	9.5~10.5	417.0~647.0
	PEG-grafted PHO	4.8~10.7	2.8~3.3	−34.5~−37.5	48.9~53.0	37.3~52.1	5.7~12.7
	PMMA-grafted PHN	5.5~47.3	2.5~5.8			0.2~0.4	379.0~621.0
	PEG-grafted PHU						

the limiting nutrient to trigger PHA production has been shown to have severe limitations, and, although sufficient, not a necessary condition for achievement of high productivity, PHA content in cells and biopolymer yields, as well for microorganisms of *Cupriavidus* genus, as for the mixed cultures.

The thermo-chemical properties of PHA have been shown to be strongly related to their structure. Thus, PHA$_{SCL}$ are mainly brittle thermoplastics with high melting and glass transition temperatures, whereas PHA$_{SCL}$ are mainly elastomers. Biocompatibility is another interesting property of PHAs. In this case, PHA$_{MCL}$ have been shown to be more biocompatible than PHA$_{SCL}$, however, they are also more hydrophobic and the hydrophobicity may be an important drawback for such types of applications. Thus, post-fermentation modifications of PHAs were also applied to modify their structure. Those modifications consist either of monomeric modifications, *i.e.* grafting, functionalization *etc.*, or surface treatment (UV, basic treatments).

Finally, the more successful enhancement of the biocompatibility properties than of the thermo-mechanical properties achieved in recent years, allows us to foresee a switch in applications, from packaging and low-added value applications to biomedical applications. This switch can also imply another one, in the field of the carbon sources used for PHA production. Indeed, until now, low-added value applications have driven the studies of cheap carbon sources, such as waste, even if in some cases (*e.g.* glycerol), the actual economical gain is doubtful. Yet biomedical applications require extreme purity of biomaterials; in these cases, the use of waste carbon sources, possibly containing micropollutants that are difficult to remove, can be an issue.

Thus, the extremely strong relationship between sources, structure and properties requires an overall consideration of the goal for the future development of the field. A global approach, considering that the carbon source influences the cost and structure of PHAs, which influence the possible applications, which in turn influence the choice of the initial carbon source used for the production of the biopolymer, then seems necessary for further improvements and the widespread use of PHAs.

References

1. A. Steinbüchel and H. E. Valentin, *FEMS Microbiol. Lett.*, 1995, **128**, 219–228.
2. K. Grage, A. C. Jahns, N. Parlane, R. Palanisamy, I. A. Rasiah, J. A. Atwood and B. H. A. Rehm, *Biomacromolecules*, 2009, **10**, 660–669.
3. J. M. Luengo, B. García, A. Sandoval, G. Naharro and E. a. R. Olivera, *Curr. Opin. Microbiol.*, 2003, **6**, 251–260.
4. R. Rai, T. Keshavarz, J. A. Roether, A. R. Boccaccini and I. Roy, *Mater. Sci. Eng., B*, 2011, **72**, 29–47.

5. D. B. Hazer, E. Kiliçay and B. Hazer, *Mater. Sci. Eng., C*, 2012, **32**, 637–647.
6. E. N. Pederson, C. W. J. McChalicher and F. Srienc, *Biomacromolecules*, 2006, 7, 1904–1911.
7. L. Tripathi, L.-P. Wu, M. Dechuan, J. Chen, Q. Wu and G.-Q. Chen, *Bioresour. Technol.*, 2013, **142**, 225–231.
8. L. Zhang, Z.-Y. Shi, Q. Wu and G.-Q. Chen, *Appl. Microbiol. Biotechnol.*, 2009, **84**, 909–916.
9. I. K. Kang, S. H. Choi, D. S. Shin and S. C. Yoon, *Int. J. Biol. Macromol.*, 2001, **28**, 205–212.
10. B. Hazer and A. Steinbüchel, *Appl. Microbiol. Biotechnol.*, 2007, **74**, 1–12.
11. S. F. Williams, D. P. Martin, D. M. Horowitz and O. P. Peoples, *Int. J. Biol. Macromol.*, 1999, **25**, 111–121.
12. R. Davis, R. Kataria, F. Cerrone, T. Woods, S. Kenny, A. O'Donovan, M. Guzik, H. Shaikh, G. Duane, V. K. Gupta, M. G. Tuohy, R. B. Padamatti, E. Casey and K. E. O'Connor, *Bioresour. Technol.*, 2013, **150**, 202–209.
13. H. E. Valentin, T. A. Mitsky, D. A. Mahadeo, M. Tran and K. J. Gruys, *Appl. Environ. Microbiol.*, 2000, **66**, 5253–5258.
14. S. J. Park, T. W. Kim, M. K. Kim, S. Y. Lee and S.-C. Lim, *Biotechnol. Adv.*, 2012, **30**, 1196–1206.
15. B. H. A. Rehm, *Biotechnol. Lett.*, 2006, **28**, 207–213.
16. I. S. Aldor and J. D. Keasling, *Curr. Opin. Biotechnol.*, 2003, **14**, 475–783.
17. M. Potter and A. Steinbuchel, *Microbiol. Monogr.*, 2006, **1**, 110–136.
18. B. S. Kim, S. C. Lee, S. Y. Lee, H. N. Chang, Y. K. Chang and S. I. Woo, *Biotechnol. Bioeng.*, 1994, **43**, 892–898.
19. S. I. Mozumber, H. De Wever, E. I. P. Volcke and L. Garcia-Gonzalez, *Process Biochem.*, 2014, **49**, 356–373.
20. H. W. Ryu, S. K. Hahn, Y. K. Chang and H. N. Chang, *Biotechnol. Bioeng.*, 1997, **55**, 28–32.
21. E. Grousseau, E. Blanchet, S. Deleris, M. Albuquerque, E. Paul and J. L. Uribelarrea, *Bioresour. Technol.*, 2013, **148**, 30–38.
22. P. Passanha, S. R. Esteves, G. Kedia, R. M. Dinsdale and A. J. Guwy, *Bioresour. Technol.*, 2013, **147**, 345–352.
23. N. Berezina, *New Biotechnol.*, 2013, **30**, 192–195.
24. E. J. Borman, M. Leissner, M. Roth, B. Beer and K. Metzner, *Appl. Microbiol. Biotechnol.*, 1998, **50**, 604–607.
25. N. Berezina, O. Talon and L. Paternostre, *Chem. Eng. Trans.*, 2008, **14**, 153–160.
26. J. G. Cruz Pradella, J. L. Ienczak, C. Romero Delgado and M. Keico Taciro, *Biotechnol. Lett.*, 2012, **34**, 1003–1007.
27. L. Shang, M. Jiang and H. N. Chang, *Biotechnol. Lett.*, 2003, **25**, 1415–1419.
28. Z. Chen, Y. Li, Q. Wen and H. Zhang, *Chemosphere*, 2011, **82**, 1209–1213.

29. H. Moralejo-Garate, T. Palmeiro-Sanchez, R. Kleerbezem, A. Mosquera-Corral, J. L. Campos and M. C. M. van Loosdrecht, *Biotechnol. Bioeng.*, 2013, **110**, 3148–3155.

30. K. Sudesh, H. Abe and Y. Doi, *Prog. Polym. Sci.*, 2000, **25**, 1503–1555.

31. B. Hazer, *Int. J. Polym. Sci.*, 2010, **2010**, 1–8.

32. G. Penloglou, C. Chatzidoukas and C. Kiparissides, *Biotechnol. Adv.*, 2012, **30**, 329–337.

33. R. Rai, T. Keshavarz, J. A. Roether, A. R. Boccaccini and I. Roy, *Mater. Sci. Eng., R.*, 2011, **72**, 29–47.

34. A. J. Anderson and E. A. Dawes, *Microbiol. Rev.*, 1990, **54**, 450–472.

35. X. H. Qu, Q. Wu, J. Liang, B. Zou and G. Q. Chen, *Biomaterials*, 2006, **27**, 2944–2950.

36. K. Zhao, Y. Deng, J. C. Chen and G. Q. Chen, *Biomaterials*, 2003, **24**, 1041–1045.

37. Y. J. Hu, X. Wei, W. Zhao, Y. S. Liu and G. Q. Chen, *Acta Biomater.*, 2009, **5**, 1115–1125.

38. Z. W. Dai, X. H. Zou and G. Q. Chen, *Biomaterials*, 2009, **30**, 3075–3083.

39. T. H. Ying, D. Ishii, A. Mahara, S. Murakami, T. Yamaoka, K. Sudesh, R. Samian, M. Fujita, M. Maeda and T. Iwata, *Biomaterials*, 2008, **29**, 1307–1317.

40. D. P. Martin and S. F. Williams, *Biochem. Eng. J.*, 2003, **16**, 97–105.

41. S. F. Williams, D. P. Martin and F. Skraly, EP 1 163 019 B1 / WO 2000/ 056376.

42. Y. Y. Shangguan, Y. W. Wang, Q. Wu and G. Q. Chen, *Biomaterials*, 2006, **27**, 2349–2357.

43. L. Zhang, Z. Zheng, J. Xi, Y. Gao, Q. Ao, Y. Gong and X. Zhang, *Eur. Polym. J.*, 2007, **43**, 2975–2986.

44. G. Q. Chen and Q. Wu, *Biomaterials*, 2005, **26**, 6565–6578.

45. X. Yang, K. Zhao and G. Q. Chen, *Biomaterials*, 2002, **23**, 1391–1397.

46. S. F. Williams, D. P. Martin, D. M. Horowitz and O. P. Peoples, *Int. J. Biol. Macromol.*, 1999, **25**, 111–121.

47. I. K. Kang, S. H. Choi, D. S. Shin and S. C. Yoon, *Int. J. Biol. Macromol.*, 2001, **28**, 205–212.

48. T. Pompe, K. Keller, G. Mothes, M. Nitschke, M. Teese, R. Zimmermann and C. Werner, *Biomaterials*, 2007, **28**, 28–37.

49. X. H. Qu, Q. Wu, J. Liang, X. Qu, S. G. Wang and G. Q. Chen, *Biomaterials*, 2005, **26**, 6991–7001.

50. F. Shen, E. Zhang and Z. Wei, *Colloids Surf., B*, 2009, **73**, 302–307.

51. Y. W. Wang, Q. Wu and G. Q. Chen, *Biomaterials*, 2003, **24**, 4621–4629.

52. X. Gao, J.-C. Chen, Q. Wu and G.-Q. Chen, *Curr. Opin. Biotechnol.*, 2011, **22**, 768–774.

53. S. Taguchi, H. Nakamura, T. Kichise, T. Tsuge, I. Yamato and Y. Doi, *Biochem. Eng. J.*, 2003, **16**, 107–113.

54. R. W. Lenz and R. H. Marchessault, *Biomacromolecules*, 2004, **6**, 1–8.

55. B. H. Rehm and A. Steinbuchel, *Int. J. Biol. Macromol.*, 1999, **25**, 3–19.

56. M. Liebergesell and A. Steinbuchel, *Eur. J. Biochem.*, 1992, **209**, 135–150.

57. M. Liebergesell, S. Rahalkar and A. Steinbuchel, *Appl. Microbiol. Biotechnol.*, 2000, **54**, 186–194.
58. Y. Satoh, N. Minamoto, K. Tajima and M. Munekata, *J. Biosci. Bioeng.*, 2002, **94**, 343–350.
59. C. W. J. McChalicher and F. Srienc, *J. Biotechnol.*, 2007, **132**, 296–302.
60. B. Laycock, P. Halley, S. Pratt, A. Werker and P. Lant, *Prog. Polym. Sci.*, 2013, **38**, 536–583.
61. B. H. Rehm, *Nat. Rev. Microbiol.*, 2010, **8**, 578–592.
62. T. Lutke-Eversloh, A. Fischer, U. Remminghorst, J. Kawada, R. H. Marchessault, A. Bogershausen, M. Kalwei, H. Eckert, R. Reichelt, S. J. Liu and A. Steinbuchel, *Nat. Mater.*, 2002, **1**, 236–240.
63. M. Tortajada, L. F. da Silva and M. A. Prieto, *Int. Microbiol.*, 2013, **16**, 1–15.
64. P. Höfer, P. Vermette and D. Groleau, *Biochem. Eng. J.*, 2011, **54**, 26–33.
65. P. Hoefer, *Front. Biosci.*, 2010, **15**, 93–121.
66. H. E. Valentin, P. A. Berger, K. J. Gruys, M. Filomena de Andrade Rodrigues, A. Steinbüchel, M. Tran and J. Asrar, *Macromolecules*, 1999, **32**, 7389–7395.
67. H. W. Ulmer, R. A. Gross, M. Posada, P. Weisbach, R. C. Fuller and R. W. Lenz, *Macromolecules*, 1994, **27**, 1675–1679.
68. A. Ballistreri, G. Montaudo, G. Impallomeni, R. W. Lenz, H. W. Ulmer and R. C. Fuller, *Macromolecules*, 1995, **28**, 3664–3671.
69. P. Hoefer, Y. J. Choi, M. J. Osborne, C. B. Miguez, P. Vermette and D. Groleau, *Microb. Cell Fact.*, 2010, **9**, 1–13.
70. Y. B. Kim, R. W. Lenz and R. C. Fuller, *J. Polym. Sci., Part A: Polym. Chem.*, 1995, **33**, 1367–1374.
71. A. Ballistreri, M. Giuffrida, S. P. P. Guglielmino, S. Carnazza, A. Ferreri and G. Impallomeni, *Int. J. Biol. Macromol.*, 2001, **29**, 107–114.
72. G. Eggink, H. van der Wal, G. N. M. Huijberts and P. de Waard, *Ind. Crops Prod.*, 1992, **1**, 157–163.
73. R. D. Ashby and T. A. Foglia, *Appl. Microbiol. Biotechnol.*, 1998, **49**, 431–437.
74. R. A. Gross, C. DeMello, R. W. Lenz, H. Brandl and R. C. Fuller, *Macromolecules*, 1989, **22**, 1106–1115.
75. H. Brandl, R. A. Gross, R. W. Lenz and R. C. Fuller, *Appl. Environ. Microbiol.*, 1988, **54**, 1977–1982.
76. K. Fritzsche, R. W. Lenz and R. C. Fuller, *Int. J. Biol. Macromol.*, 1990, **12**, 92–101.
77. M. H. Choi and S. C. Yoon, *Appl. Environ. Microbiol.*, 1994, **60**, 3245–3254.
78. K. M. Tobin and K. E. O'Connor, *FEMS Microbiol. Lett.*, 2005, **253**, 111–118.
79. H. Ritter and A. G. von Spee, *Macromol. Chem. Phys.*, 1994, **195**, 1665–1672.
80. Y. B. Kim, Y. H. Rhee, S.-H. Han, G. S. Heo and J. S. Kim, *Macromolecules*, 1996, **29**, 3432–3435.
81. T. Imamura, T. Kenmoku, T. Honma, S. Kobayashi and T. Yano, *Int. J. Biol. Macromol.*, 2001, **29**, 295–301.

82. S. M. Arostegui, M. A. Aponte, E. Diaz and E. Schroder, *Macromolecules*, 1999, **32**, 2889–2895.
83. R. H. Marchessault, C. J. Monasterios, F. G. Morin and P. R. Sundararajan, *Int. J. Biol. Macromol.*, 1990, **12**, 158–165.
84. H. Preusting, A. Nijenhuis and B. Witholt, *Macromolecules*, 1990, **23**, 4220–4224.
85. R. J. Sánchez, J. Schripsema, L. F. da Silva, M. K. Taciro, J. G. C. Pradella and J. G. C. Gomez, *Eur. Polym. J.*, 2003, **39**, 1385–1394.
86. S. Domenek, V. Langlois and E. Renard, *Polym. Degrad. Stab.*, 2007, **92**, 1384–1392.
87. Y. Kim do, H. W. Kim, M. G. Chung and Y. H. Rhee, *J. Microbiol.*, 2007, **45**, 87–97.
88. P. Lemechko, J. Ramier, D. L. Versace, J. Guezennec, C. Simon-Colin, P. Albanese, E. Renard and V. Langlois, *React. Funct. Polym.*, 2013, **73**, 237–243.
89. P. Lemechko, E. Renard, J. Guezennec, C. Simon-Colin and V. Langlois, *React. Funct. Polym.*, 2012, **72**, 487–494.
90. H. Kukula, H. Schlaad, M. Antonietti and S. Forster, *J. Am. Chem. Soc.*, 2002, **124**, 1658–1663.
91. J. Li, X. Ni, X. Li, N. K. Tan, C. T. Lim, S. Ramakrishna and K. W. Leong, *Langmuir*, 2005, **21**, 8681–8685.
92. A. P. Andrade, B. Witholt, R. Hany, T. Egli and Z. Li, *Macromolecules*, 2001, **35**, 684–689.
93. T. D. Hirt, P. Neuenschwander and U. W. Suter, *Macromol. Chem. Phys.*, 1996, **197**, 1609–1614.
94. K. J. Zhu, S. Bihai and Y. Shilin, *J. Polym. Sci., Part A: Polym. Chem.*, 1989, **27**, 2151–2159.
95. F. Ravenelle and R. H. Marchessault, *Biomacromolecules*, 2003, **4**, 856–858.
96. G. Adamus, W. Sikorska, H. Janeczek, M. Kwiecien, M. Sobota and M. Kowalczuk, *Eur. Polym. J.*, 2012, **48**, 621–631.
97. Z. Jedlinski, M. Kowalczuk, G. Adamus, W. Sikorska and J. Rydz, *Int. J. Biol. Macromol.*, 1999, **25**, 247–253.
98. M. S. Reeve, S. P. McCarthy and R. A. Gross, *Macromolecules*, 1993, **26**, 888–894.
99. M. Yalpani, R. H. Marchessault, F. G. Morin and C. J. Monasterios, *Macromolecules*, 1991, **24**, 6046–6049.
100. H. Arslan, B. Hazer and S. C. Yoon, *J. Appl. Polym. Sci.*, 2007, **103**, 81–89.
101. X. Li, X. J. Loh, K. Wang, C. He and J. Li, *Biomacromolecules*, 2005, **6**, 2740–2747.
102. J. Hao and X. Deng, *Polymer*, 2001, **42**, 4091–4097.
103. H. W. Kim, C. W. Chung and Y. H. Rhee, *Int. J. Biol. Macromol.*, 2005, **35**, 47–53.
104. A. H. Arkin, B. Hazer and M. Borcakli, *Macromolecules*, 2000, **33**, 9160–9160.
105. A. H. Arkin and B. Hazer, *Biomacromolecules*, 2002, **3**, 1327–1335.

106. K. D. Gagnon, R. W. Lenz, R. J. Farris and R. C. Fuller, *Polymer*, 1994, **35**, 4358–4367.

107. K. D. Gagnon, R. W. Lenz, R. J. Farris and R. C. Fuller, *Polymer*, 1994, **35**, 4368–4375.

108. A. Dufresne, L. Reche, R. H. Marchessault and M. Lacroix, *Int. J. Biol. Macromol.*, 2001, **29**, 73–82.

109. M.-M. Bear, M.-A. Leboucher-Durand, V. Langlois, R. W. Lenz, S. Goodwin and P. Guérin, *React. Funct. Polym.*, 1997, **34**, 65–77.

110. W. H. Park, R. W. Lenz and S. Goodwin, *Macromolecules*, 1998, **31**, 1480–1486.

111. M. Y. Lee, W. H. Park and R. W. Lenz, *Polymer*, 2000, **41**, 1703–1709.

112. M. S. Eroğlu, B. Hazer, T. Ozturk and T. Caykara, *J. Appl. Polym. Sci.*, 2005, **97**, 2132–2139.

113. R. Hany, C. Böhlen, T. Geiger, R. Hartmann, J. Kawada, M. Schmid, M. Zinn and R. H. Marchessault, *Macromolecules*, 2003, **37**, 385–389.

114. H. Erduranli, B. Hazer and M. Borcakli, *Macromol. Symp.*, 2008, **269**, 161–169.

115. M. Constantin, C. I. Simionescu, A. Carpov, E. Samain and H. Driguez, *Macromol. Rapid Commun.*, 1999, **20**, 91–94.

116. I. V. Spoljaric, M. Lopar, M. Koller, A. Salerno, A. Reiterer and P. Horvat, *J. Biotechnol.*, 2013, **168**, 625–635.

117. J. M. B. T. Cavalheiro, C. M. D. Almeida, C. Grandfils and M. M. R. Fonseca, *Process Biochem.*, 2009, **44**, 509–515.

118. I. V. Spoljaric, M. Lopar, M. Koller, A. Muhr, A. Salerno, A. Reiterer, K. Malli, H. Angerer, K. Strohmeier, S. Schober, M. Mittelbach and P. Horvat, *Bioresour. Technol.*, 2013, **133**, 482–494.

119. M. A. K. M. Zahari, M. R. Zakaria, H. Ariffin, M. N. Mokhtar, J. Salihon, Y. Shirai and M. A. Hassan, *Bioresour. Technol.*, 2012, **110**, 566–571.

120. M. A. K. M. Zahari, H. Ariffin, M. N. Mokhtar, J. Salihon, Y. Shirai and M. A. Hassan, *J. Biomed. Biotechnol.*, 2012, **2012**, 1–8.

121. K. M. Lee and D. F. Gilmore, *Process Biochem.*, 2005, **40**, 229–246.

122. S. Obruca, O. Snajdar, Z. Svoboda and I. Marova, *World J. Microbiol. Biotechnol.*, 2013, **29**, 2417–2428.

CHAPTER 3

Recovery and Extraction of Polyhydroxyalkanoates (PHAs)

MITRA MOHAMMADI*[a] AND
MANSOUR GHAFFARI-MOGHADDAM[b]

[a] Department of Environmental Science, Kheradgerayan Motahar Institute of Higher Education, Mashhad, Iran; [b] Department of Chemistry, Faculty of Science, University of Zabol, Zabol, Iran
*Email: m.mohammdi@motahar.ac.ir; mitramohammadi@gmail.com

3.1 Introduction

The cost of PHA production is mainly affected by downstream processing and therefore, the development of PHA extraction methods is required to make the overall process much simpler and cheaper.[1] Improved bio-separation systems are essential for biotechnology as separation is the limiting parameter for the success of biological processes.[2] The recovery system may affect the amount of product recovery, the convenience of the subsequent purification steps and the quality of the final product. Cell separation from fermentation broth is the preliminary step of the recovery method.[3] Figure 3.1 shows the morphology of PHA granules in a bacteria cell when observed using a transmission electron microscope (TEM). In order to recover the PHA granules, it is necessary to rupture the bacterial cell and remove the protein layer that coats the PHA granules. Alternatively, the PHA has to be selectively dissolved in a suitable solvent. Generally, two methods are usually utilized for the recovery and purification of PHA from a cell biomass, which include PHA solubilization or non-PHA cell mass (NPCM) dissolution. In the first one, PHA granules are dissolved in a suitable organic

RSC Green Chemistry No. 30
Polyhydroxyalkanoate (PHA) Based Blends, Composites and Nanocomposites
Edited by Ipsita Roy and Visakh P M
© The Royal Society of Chemistry 2015
Published by the Royal Society of Chemistry, www.rsc.org

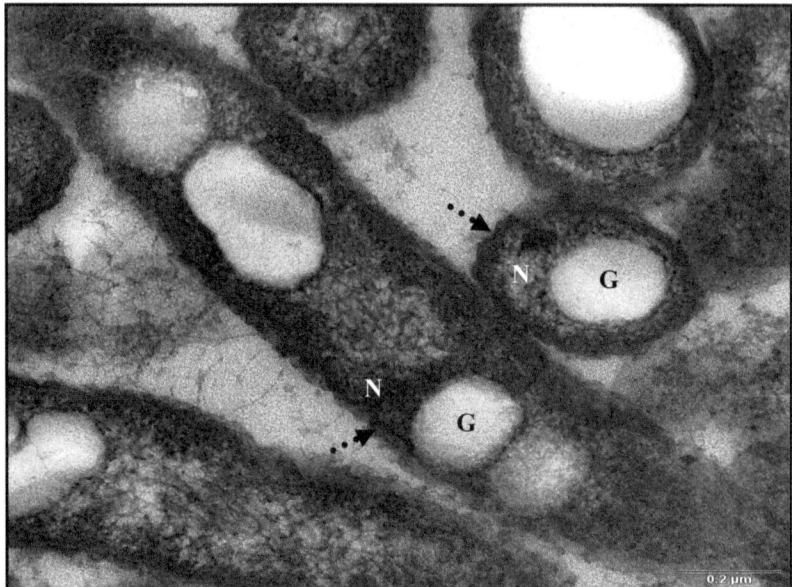

Figure 3.1 The morphology of PHA granules in a bacteria cell. The round dot arrow indicates the cell wall; N = NPCM; G = PHA granule.

solvent and extracted from the NPCM. However, PHA polymers remain in the form of a solid during the second process while NPCM is digested and/or dissolved by agents. Finally, the obtained solid and liquid phases are separated using filtration or centrifugation. A useful classification of different recovery methods that are available to extract the PHA from the cell is given in Figure 3.2.[3,4]

Some of the most successful laboratory cell disruption techniques have no possibility of commercialization. Selecting the best recovery method depends on factors such as types of cells and their history, sample volume, reaction time, possible scale-up potential, effect on downstream purification processes and economics of disruption. It has been reported that the efficiency of a recovery process is dependent on the type of cell. Besides, cell growth and their storage history can be affected by the destructibility of cells. The bacteria cells in log phase growth tend to produce thinner cell walls, which are easier to disrupt. This and other conditions that can influence microbial cell destructibility are carbon source, micronutrients and media richness, phase of growth (batch fermentation), retention time (continues fermentation) and strain of microorganism. Moreover, the operating and energy requirements, availability and cost of the disruption equipment are other important parameters.[5]

More details about the most common techniques of PHA recovery, which are classified as chemical, biological, mechanical and physical methods as independent systems or in combination with other processes, will be presented in the following sections.

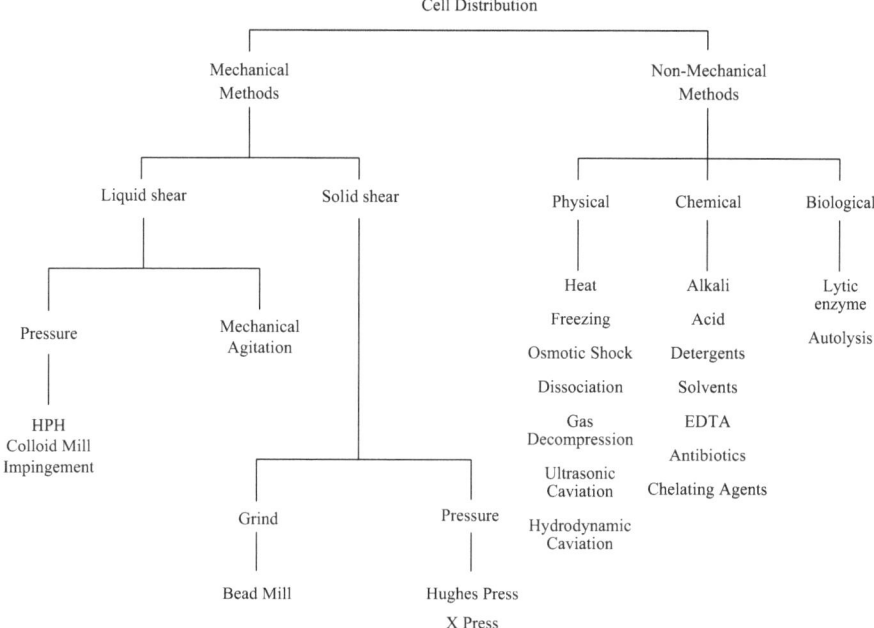

Figure 3.2   Classification of cell disruption techniques.[3,4]

3.2   Recovery of PHAs

3.2.1   Chemical Methods

3.2.1.1   Solvent Extraction

The majority of PHA recovery methods is performed using a solvent extraction process mainly by chloroform and methanol. Modifying the cell wall permeability and then PHA dissolution in the solvent are its mechanisms for PHA extraction. The separation of PHA from the solvent was carried out using solvent evaporation or polymer precipitation in a non-solvent material.[6] Solvent extraction does not degrade the polymer compared to some other recovery methods. It can be valuable for some medical applications due to the elimination of endotoxin, which can be found in Gram negative bacteria.[7] This process requires cell pretreatment steps such as grinding and spray-drying to improve the solvent extractability. Therefore, this method is considered as an economically unfavorable process due to several steps and the relatively high cost of solvents. Austrian company Chemie Linz and Imperial Chemical Industries (ICI) in the United Kingdom initially used solvent extraction for PHA recovery, however it was costly. As another weakness, the solvents can alter the natural morphology of PHA granules.[8]

It has been reported that the extracted polymer solution with more than 5% (w/v) PHA is very viscous and the removal of cell debris is difficult. With no consideration of solvent recycling, the large amount of solvent required for PHA extraction makes it a costly process.[9] Consequently, PHA solubilizing into an immiscible solvent in water was explored at a high temperature (above 120 °C). Then, cold water is added to extract the PHA, although the solvent can be recycled many times before being distilled.[10]

The hazardous impacts of solvents for operators and also the environment are considered as another problem related to their usage. This process is suitable for laboratory studies only with limited success in pilot-plant and large scale processing.[11] Recently, Metabolix (USA) has been using organic solvents for PHA recovery, which could be a promising process for reducing PHA extraction costs.[12]

3.2.1.2 *Chemical Digestion*

3.2.1.2.1 Digestion by Surfactants. Surfactants such as anionic sodium dodecyl sulfate (SDS) disrupt cells by incorporating themselves into the lipid membrane. The volume of the cell membrane is increased until it is saturated by further addition of surfactant. The cell envelope breaks once it is saturated and produces micelles of surfactant and membrane phospholipids. This leads the PHA to be released to the solution surrounded by cell debris.[13] Solubilization of NPCM such as proteins is another function of the surfactant in the recovery process.[6] Breakage of the cell membrane by a surfactant without degrading polymer granules is considered as its advantage. However, a high degree of PHA purity cannot be obtained using surfactant alone and its combination with other agents, such as hypochlorite and sodium hydroxide, is required. Furthermore, the surfactant is more costly compared with acids or alkalis and also causes problems in wastewater treatment and reuse.[12]

3.2.1.2.2 Digestion by Sodium Hypochlorite. Another popular recovery method is usage of sodium hypochlorite, which solubilizes NPCM and leaves the PHA intact. Then, the PHA can be separated from the solution by centrifugation. Nevertheless, it degrades the PHA severely, leading to a 50% molecular weight reduction. This is because sodium hypochlorite is a strong oxidant, and a significant amount of chlorine remains in the recovered PHA.[14] Generally, native amorphous PHA granules are relatively susceptible to alkaline saponification and are rapidly decomposed to monomers and oligomers as soluble products.[12] Addition of an antioxidant such as sodium bisulfite can avoid the degradation of molecular weight by sodium hypochlorite. Moreover, this method was not economically efficient.[15]

3.2.1.2.3 Digestion by Sodium Hypochlorite and Chloroform. A new PHA recovery process was developed to obtain the benefits of both digestion methods using sodium hypochlorite and chloroform as solvent extraction. The combined method makes three separate phases, including hypochlorite solution at the upper phase, NPCM and undisrupted cells at the middle phase and a chloroform phase containing PHA. The polymer is then recovered by precipitation in a non-solvent and filtration. The molecular weight reduction due to polymer degradation is significantly reduced using this process. It is due to the immediate dissolution of PHA released after hypochlorite treatment in chloroform. However, the use of sodium hypochlorite with a large amount of chloroform make this process economically unfavourable and environmental hazardous.[16]

3.2.1.2.4 Surfactant–Hypochlorite Digestion. The surfactant–hypochlorite digestion method is a cost-effective process with a limited effect on PHA degradation. It has been evaluated that US$ 5.58/kg of P(3HB) is produced for the annual production of 2850 tonnes of purified P(3HB) using the surfactant–hypochlorite digestion method from *C. necator*. However, it has a lower price compared to the chloroform–hypochlorite system. The cost of P(3HB) declined to US$ 4.75/kg due to an increase in the production scale to one million tonnes per year.[17]

3.2.1.2.5 Surfactant–Chelate Digestion. The releasing rate of PHA is improved by the addition of a chelate to the surfactant. Chelate addition can destabilize the outer membrane of Gram negative bacteria such as *C. necator* by forming complexes with their divalent cations such as Ca^{2+} and Mg^{2+}. Furthermore, the weakness in the inner membrane occurs by modifying the outer membrane.[6] Suitable operation, high quality of product and low environmental pollution are the advantages of this method while wastewater production during the recovery is its drawback. A continuous process with the application of recycled wastewater is required which leads to a higher cost.[18]

3.2.1.2.6 Selective Dissolution of NPCM. Selective dissolution of cell mass is a new and effective system for PHA recovery and purification, which is more economical than other techniques. It was estimated that its usage on a large scale can reduce the chemical cost of PHA recovery by 90%. Besides, selective dissolution of NPCM by protons in aqueous solution and PHA crystallization are known as its mechanisms for PHA recovery. This treatment is also followed by a final decolorization in a bleaching solution.[19]

3.2.1.2.7 Digestion by Sodium Hydroxide and Potassium Hydroxide. Much effort is carried out to use less expensive chemicals for PHA recovery *via* NPCM digestion.[20] Amongst these chemicals, sodium hydroxide (NaOH) and potassium hydroxide (KOH) were found to be good candidates for PHA recovery, since they are cheap and efficient, which leads to high PHA yield and purity. Also, the amount of NaOH and KOH used is much less than other chemicals to obtain similar recovery efficiencies. Purified PHA obtained by the NaOH method is more suitable for biomedical applications since it can decrease the amount of endotoxin in the polymer.[7] However, it can alter the physical and mechanical properties of PHA, which is due to the considerable hydrolysis of the polymer and its molecular weight reduction.[12] PHA with 91–92% purity and a yield of 90–93% was obtained from recombinant *E. coli* harbouring the PHA gene of *C. necator* using 0.1 M NaOH or KOH solution at 30 °C for 1 h.[21] However, PHA loss and molecular weight reduction were observed in a strong alkaline solution.[22] Moreover, Choi and Lee[21] found that PHA could be most efficiently recovered from recombinant E. coli by NaOH digestion since the cells become fragile as a result of high PHA accumulation. It was estimated that the overall PHA production cost coupled with the NaOH recovery method is US\$ 3.66/kg P(3HB), which was 25% less than the surfactant–hypochlorite digestion method.[17]

3.2.2   Biological Methods

3.2.2.1   Enzymatic Digestion

Enzymatic digestion is considered as an alternative method for solvent extraction, which is gentle (100% biological) and selective. Some enzymes, such as proteolytic enzymes, have a high ability for protein dissolution with minor effects on PHA degradation. Sterilization of the fermentation broth for a short period of time is required to provide a heat shock during enzymatic digestion. Heat pretreatment can break the cells, and solubilize and denature the polynucleic acids, which avoids an increase in medium viscosity. Then, the solubilization of NPCM can be carried out during different enzymatic treatments.[12]

Digesting the denatured nucleic acids and proteins is carried out using alcalase as a protease enzyme while lysozyme is suitable for degrading the peptidoglycan wall around the PHA granule. A lower amount of lysozyme should be sufficient since the viscosity of the solution is increased as a result of peptidoglycan wall breakage.[23] The enzymatic recovery of P(3HB) obtained from *C. necator* using 2% of bromelain at 50 °C and pH 9.0 led to 88.8% purity. Moreover, 90% purity was achieved during enzymatic digestion using pancreatin, which is three times cheaper than bromelain. Generally, the price of PHA obtained by this procedure was US\$ 84/kg P(3HB), which is high compared to other systems.[4]

3.2.3   Mechanical Methods

A few mechanical methods have also been developed as a supplement to the systems described earlier or as independent systems, which are widely used to recover intracellular PHA.[24] This field involves either solid shear (*e.g.* bead milling, extrusion of frozen cells) or liquid shear (*e.g.* high pressure homogenization or HPH). Mechanical cell disruption methods, such as HPH and bead mills, are more economical and have been widely used on a large scale.[3] However, they have several disadvantages such as generation of high temperature in cell suspensions and high shear stress, they require specialized equipment for disruption, high capital investment and energy costs, they are time-consuming, and tend towards cross-contamination and production of fine cell debris.[25]

3.2.3.1   Bead Mill

A bead mill contains a vertical cylindrical grinding chamber including a concentric cylinder with a variable speed rotor for agitation. Cell slurries go into the mills, increase the annular gap between the rotor and the stator and exit near the top. Heat is generated during the procedure, which can be controlled by circulating cooling water in the jacket around the grinding chamber. The bead mill has no difficulties in scale up since its disruption is constant and not dependent on the biomass concentration. Besides, it was reported that the disruption rate of bead mills is affected by the bead loading, however, it is not dependent on the diameter of grinding beads. Several passes through the bead mill is required to achieve high PHA purity and recovery yield, which leads to an increase in the recovery cost.[24]

3.2.3.2   High Pressure Homogenization

HPH is one of the most widely known methods for large scale cell disruption.[25] Cell disruption using HPH occurs at ambient temperature (25 °C), which is sustained by immersing the disruption chamber and the exit lines in ice during operation. The performance of the homogenizer depends on biomass concentration, pH, temperature, pressure, chemical composition of the sample, composition of medium and microbial species.[24] Medium composition affects the efficiency of HPH mainly due to the interaction of medium components with microorganisms. Pressure and temperature are important factors that control the performance of homogenization for cell disruption. Microbial strain as a function of the organism's nature and its culture conditions has a superior effect on the efficacy of HPH treatment compared to pressure.[26] It has been found that the disruption of Gram negative bacteria is easier than Gram-positive bacteria, which is mainly due to their cell wall composition, size, shape and growth phase.[25] Moreover, blockages and little micronization are some of the difficulties that can be faced during this process. Tamer and Moo-Young[27] found that the homogenizer was not effective for disruption of *A. latus* at low biomass concentrations while its performance enhanced with a further increase of concentration until blockages made operation impractical.

3.2.4   Physical Methods

3.2.4.1   Ultrasonication

Sonication is a successful method which has been extensively used for cell disruption as a result of its speed, simplicity, cleanliness and capability to lyse different kinds of cells.[28] Ultrasonic vibration is sound with a frequency higher than 15–20 kHz, which is inaudible to the human ear. It leads to inactivation and disruption of microbial cells in suspension. The application of this process for large scale cell disruption is not practiced, which is due to heat generation during the process and the high cost of a good temperature controller. In addition, ultrasonication causes chemical effects such as formation of free radicals that damage some of the molecules of interest. Production of very fine cell debris happens throughout the process that can cause difficulties in subsequent processing. The problem with transmitting sufficient power to the large volumes of cell material has been considered as another disadvantage.[3] Cell membrane disruption by sonication mainly occurs by ultrasound induced cavitation.[29] At satisfactorily high acoustic power inputs, microbubbles appear at various nucleation sites in the fluid. These bubbles grow within the rarefaction phase of the sound wave and then are compressed to a minimum radius in the compression phase. It leads to the release of a violent shock wave which spreads through the medium.

The formation, growth and collapse of gas and vapour bubbles due to the action of intense sound waves is expressed as cavitation. A large amount of sonic energy is changed to elastic waves as mechanical energy during the collapse phase, which can produce great shear forces with the ability to disrupt membranes.[3] The studies of Wase and Patel[29] using *Saccharomyces cerevisiae*, *Bacillus cereus* and *E. coli* conclusively showed that the performance of ultrasonication in the disruption of a cell wall is dependent on the cell volume and growth phase.

3.2.4.2 Osmotic Shock

It has been found that water can leave cells when it is attracted by high salt concentrations, which makes the cell shrivel and dehydrate. Osmotic pressure is considered either the quick dilution of a solution with high osmotic pressure or the rapid resuspension of cells in a solution with a high salt concentration.[27] This method is not so effective for microbial cell disruption due to the structure of the cell wall. Firstly, the cell walls have to be weakened in order to achieve the release of cytoplasmic components. It can release some of the proteins from Gram-negative bacteria cells.[25] Furthermore, there is a need to combine this method with another process, such as alkaline pretreatment, to improve its effectiveness. Whereas this process is appropriate for disruption of fragile cells, the contamination of the product by high concentrations of salt or another agent is objectionable.[27,30]

3.2.4.3 Freezing

Freezing and thawing of bacterial cells cause cell disruption throughout the formation and a consequent melting of ice crystals. Gradual freezing produces larger crystals that cause more severe damage to the cells. Besides, cell breakage is further improved by simultaneous grinding. Application of this method has been limited to a small scale because of its low yields and high energy requirement.[25]

3.2.5 Other PHA Recovery Methods

Recently, new methods such as supercritical fluids, spontaneous liberation of PHA, dissolved air flotation, and air classification, are being investigated and probably promise to have much more success. Improvement of these new extraction and purification methods should cause an optimal recovery of PHA, with a high purity and recovery level at a low production cost.

3.2.5.1 Supercritical Fluids

Supercritical fluids have unique physicochemical properties such as high densities and low viscosities that make them suitable as extraction solvents. For that purpose, CO_2 is most widely used because of its low toxicity and reactivity, moderate critical temperature and pressure (31 °C and 73 atm), availability, low cost and non-flammability. However, this method still seems to be costly compared with other techniques.[31]

3.2.5.2 Dissolved-air Flotation

This method continues with some of the enzyme pretreatments such as lysozymes and novozymes. The function of this process is based on the fact that cell debris and PHA polymers have an iso-electric point at a specific pH thereby selective aggregation and flotation can occur. Particle properties, such as particle size, hydrophobicity and surface charge, have a significant effect on the performance of the flotation process. However, this system does not use chemicals but it needs some subsequent flotation steps which affect the economics of the overall process.[32]

3.2.5.3 Air Classification

Air classification begins with an ultrasonic sonicator to treat the cells and create a suspension of polymer granules. The suspension is freeze dried and pulverized subsequently using a fluid energy mill. Then the grinded sample is air classified to make a 38% fine fraction and a 62% coarse fraction. The fine fraction is exposed to chloroform extraction followed by methanol precipitation to form PHA particles. This method leads to high purity but low recovery yield.[33]

3.2.5.4 Spontaneous Liberation

Bacterial strains such as *E. coli* which are capable of spontaneously liberating PHA from cells are favourably used in this system. Currently, spontaneous liberation is drawing much attention since no chemicals are employed during the procedure. Interestingly, it has been found that by modifying the initial inoculum size and the medium composition of *E. coli,* PHA granules could be released from the cell due to its accumulation and subsequent lyses of the cell wall.[34]

3.2.6 Combined Methods

Combinations of methods such as chemical and physical processes can sometimes produce acceptable results whilst one method alone fails.[5] It has been reported that chemical pretreatments increase the sensitivity of bacteria to disruption. They allow equal disruption to be obtained at lower operating pressures or fewer passes during the physical process. Alkaline pH pretreatment at pH 10.5 caused a 37.5% increment in soluble protein release following homogenization. An increase of about 30% in soluble protein release was found following the addition of 0.137 M monovalent cations (Na^+ or K^+) at 60 °C. Maximum cell breakage can thus be completed at reduced energy inputs.[35] Harrison *et al*[35] also found that the disruption efficiency of HPH under alkaline pretreatment of biomass (pH 10.5, 7 °C, ≤1 min) substantially improved. However, a minimum of two passes were still needed for complete protein release.[35] In a study done by Tamer and Moo-Young,[27] it was shown that most of the protein was released from *A. latus* within three passes of bead mill using an optimum hydroxide concentration of 0.4 kg kg^{-1} as pretreatment, while at least ten passes of a bead mill is needed for the same level of disruption in untreated cells. Moreover, molecular weight reduction was observed due to alkaline hydrolysis.[27] Besides, a combination of heat pretreatment at 50 °C followed by the 1 M NaOH method for 5 min of reaction time led to PHA purity and recovery yield of 99% and 96%, respectively.[22]

Other inventors indicated that it is promising to efficiently extract highly pure PHA by the addition of an alkali to the fermentation broth including a PHA while stirring. This was followed by a physical disruption treatment at relatively low temperature and an enzyme and/or surfactant in an aqueous suspension or wet state with further washing of PHA with a hydrophilic solvent and/or water. Furthermore, it possible to enlarge the diameter of a PHA using its suspension in a hydrophilic solvent and/or water followed by its agglomerating during stirring at a temperature equal to or below the boiling point of the suspension. These techniques are able to extract highly pure PHA with over 90% purity and recovery yield and limited reduction in molecular weight. A PHA with more improved purity may be isolated by washing it with hydrophilic solvent and/or water.[36]

Moreover, the morphology of the resulting PHA may be affected by either addition of water to the fermentation broth containing PHA or the fermentation broth to the water. In the former way, the fermentation broth and the water can be mixed together using a propeller, turbine, high shear and mixtures thereof. The resulting morphology of the PHA is typically particles, fibrous and powder where a propeller, turbine and high shear are used, respectively. It is assumed that mild mixing with both radial and vertical flows should facilitate the precipitation process. The morphology of the precipitated polymer is affected by the water addition order, rate, temperature and ratio in combination with a mixer, such as a propeller, turbine and high shear homogenizer. However, a PHA with fibrous or agglomerate morphology can be obtained with or without mixing in the latter way. Turbine mixing or the application of a high shear homogenizer causes a fibrous or agglomerate state, respectively. The rate of water addition and the mixer speed were found as key factors that influenced the morphology of the PHA. Besides, these techniques are able to enhance the purity of PHA using its precipitation in water.[37]

Furthermore, Page and Cornish[38] reported that P(3HB)'s dry weight was increased from 80% to 86% by mixing or boiling the cells in distilled water. Thus, it was suggested that cell lyses can occur in distilled water during the cell slurry preparation.

3.3 Characterization of Recovered PHAs

PHAs are biopolymers synthesized by various bacteria from renewable sources and have drawn much interest from academic and industrial communities for their unique properties of biodegradability and biocompatibility.[39] PHAs are rapidly developing into an industrial value chain ranging from packaging materials, fine chemicals to biomedical applications.[40] More than 150 PHAs have been detected through genetic and metabolic engineering, control of the cultivation conditions and feeding of suitable precursor substrates. The PHAs' properties are related to their monomer composition and can be varied in a broad range without losing biodegradability.[11] According to the monomer structures, PHAs are divided into SCL PHA commonly consisting of P(3HB); MCL PHA including poly-hydroxyhexanoate [P(3HHx)] and copolymers of SCL-MCL PHA such as P(3HB-*co*-3HV) and P(3HB-*co*-3HHx).[40] Nevertheless, due to the high production cost of PHAs, semi-commercial production has been realized only for P(3HB), P(3HB-*co*-3HV) and P(3HB-*co*-3HHx), which are restricted to several medical and technical applications.[41]

P(3HB) is known to be brittle and stiff and has a high degree of crystallinity (75–85%).[42,43] This homopolymer has shown some potential for packaging, drug delivery and medical implant applications.[44,45] However, P(3HB) homopolymer has narrow processing temperatures due to degradation and brittleness at room temperature which limits its commercial applications.[1,46] In order to expand the quality and potential applications of P(3HB), much effort has been directed to the discovery and development of new monomers and polymer compositions. P(3HV) and P(3HHx) are kinds of comonomers that have been incorporated into P(3HB) throughout the fermentation process to improve the polymer characteristics.[42,43] The resulting copolymers of P(3HB-*co*-3HV) and P(3HB-*co*-3HHx) with lower melting point, higher melt stability and less brittleness have broadened its applications.[1,46]

The success of PHA as a viable option for petrochemical-based plastics will depend upon the design and performance of efficient and selective means of PHA production and recovery.[47] Thus, further investigation on mixed cultures, recombinant microbial strains, cheap carbon substrates and efficient fermentations has allowed the production of significant quantities of PHA which can significantly decrease the production cost.[1,20,48] However, there are difficulties in obtaining efficient and cost-effective PHA recovery from bacteria cells at a valuable level of quality and purity. A commercial recovery system with a simple, efficient and economical procedure will probably focus on a non-solvent extraction-based recovery amongst a variety of PHA recovery methods. In addition, the tolerance of the final product to the conditions employed is an important criterion for the selection of a PHA recovery process and the PHA properties have to be considered for the development of downstream processes. It has been reported that polymer molecular weight is an important property as it determines many physical and mechanical characteristics such as glass transition temperatures, stiffness, strength, elasticity, toughness, and viscosity. If the PHA molecular weight is too low, the transition temperatures and the mechanical properties will usually decrease, which is not suitable for any useful commercial applications.[49] Depending on the chemical composition and the molecular weight, PHAs can be converted to a range of finished products including films and sheets; moulded articles; fibres; elastics; laminates and coated articles; non-woven fabrics; synthetic paper products and foams. PHAs are valuable for injection moulding and melt blowing at low comonomer content and molecular weight. At medium molecular weight, the material is suitable for melt spun fibres. With higher comonomer content and medium molecular weight (600 000 Da), applications consist of melt resins and cast films. Blown films and blow moulding require at least 10% of comonomer content and high molecular weight (700 000 Da). Above 15% comonomer, the PHAs are softer and more elastic, finding applications in adhesives and elastomeric films.[50] Hence, the challenge in the recovery process should be the maintenance of the original molecular weights while not compromising the degree of purity for various applications.[51] However, severe degradation of polymer molecular weight was reported during PHA extraction of *C. necator* using sodium hypochlorite treatment.[16] Ramsay *et al*[13] also indicated that the molecular weight of PHA recovered from *C. necator* by surfactant–hypochlorite was about 60% of the original molecular weight. Besides, surfactant–chelate treatment for 10 min at 50 °C caused a decreasing in the molecular weight from 402 000 to 316 000 Da.[6] If the process parameters are not controlled properly, a significant reduction of molecular weight can be observed during PHA extraction using selective dissolution of NPCM by proton.[19] Therefore, as the extraction process may considerably affect the polymer features, it is necessary to characterize and compare the extracted and non-extracted polymer properties to assess the feasibility of the developed recovery methods on PHA extraction and address the possible market demand and intended applications. Accordingly, the native PHA granules of bacterial cells before and after the recovery process can be observed under transmission electron microscopy (TEM) and scanning electron microscopy (SEM).[52] The PHAs are chemically characterized and quantified by nuclear magnetic resonance (NMR) spectroscopy.[53] Furthermore, the molecular weight of extracted and non-extracted copolymers is examined using gel permeation chromatography (GPC).[54] The morphological state of PHA granules synthesized by bacterial cells can be determined using liquid nitrogen cooled differential scanning calorimetry (DSC). Data for enthalpy of fusion and enthalpy of crystallinity are collected during the second heating cycle.[55]

3.4 Conclusion

The main obstacle to the commercial production and application of PHAs in consumer products is their high cost compared to synthetic plastics.[1] The recovery of PHA as an intracellular product significantly affects the overall economics, and therefore, developing a clean, simple and efficient process for PHA extraction from source materials at a useful level of quality and purity is a remarkable proposal.[56] Generally, it could be concluded that the extraction and purification of PHA granules from a cell biomass is a challenging task especially when one considers the use of environmentally hazardous chemicals as an unacceptable option in the production of eco-friendly materials. Various methods such as chemical, biological, mechanical and physical treatments have been developed successfully for PHA recovery from bacteria cells on a small scale. A useful comparison of the various types of methods that have been tested for the extraction of PHA granules from microbial cell biomass including their advantages and disadvantages is given in Table 3.1. Moreover, Table 3.2 shows the summary of purity, recovery yield and molecular weight of PHA recovered from various organisms by non-halogenated solvent processes, as reported in the literature. It is seen that the developed halogen-free methods could be considered as green alternative technologies for PHA recovery, which are able to eliminate the usage of harsh organic solvents and their negative impact on the environment.

Some of the efficient laboratory techniques have no opportunity for commercialization due to several factors such as high energy requirements, low accessibility and high cost of the equipment. Hence, significant parameters affecting the performance of recovery processes have to be determined for their success on an industrial scale. The selection of suitable PHA extraction methods depends on several process parameters such as concentration of chemicals, reaction time, recovery temperature, pH, *etc.* In addition, the choice of recovery system, PHA-producing bacteria, composition of the growth medium, presence of certain chemical compounds in the environment, the intracellular PHA content, length of PHA granules, cell wall structure and economics of process are considered as the most important external factors affecting the extraction process.[25,64] Basically, the

Table 3.1 Comparison of PHA recovery methods.[12]

Recovery method	Advantages	Disadvantages
Solvent extraction	Elimination of endotoxine, high purity, no polymer degradation	Breaks PHA granules' morphology, environmentally hazardous, high price, low recovery, difficult to scale up
Surfactant digestion	Treatment of high cell densities, no polymer degradation	Low purity, wastewater treatment needed, high cost compared to alkali and acid treatment
Sodium hypochlorite digestion	High purity	Degradation of the polymer, not economic
Sodium hypochlorite and chloroform digestion	Low polymer degradation, high purity	High quantity of solvent needed, high cost, not environmentally friendly
Surfactant–hypochlorite digestion	Limited degradation, low operating cost	—
Surfactant–chelate digestion	High purity, high quality of product, low environmental pollution	Waste disposal problem
Selective dissolution of NPCM	High yield, high purity, low operating costs	—
Sodium hydroxide digestion	Economic, less toxic, high purity, high yield, decrease of endotoxine	Hydrolysis of polymer, decrease of molecular weight
Enzymatic digestion	Selective, good recovery	High cost of enzymes
Bead mill	Consistent and predictable results, easy to scale up, no chemicals used	Requires several passes
High pressure homogenization	Easy to scale up, no chemicals used	Poor disruption rate for low biomass levels, low micronization, blockage problem, produces Heat
Ultrasonication	Good for lab scale, cleanliness, speed	High energy requirement, heat appearance, not suitable for large scale, damages the polymer, micronization
Osmotic shock	—	Contamination of final product, not effective
Freezing	—	Costly, high energy consumption, slow, low yield
Supercritical CO_2	Low cost, low toxicity	Low recovery
Dissolved air flotation	No chemicals used	Requires several consecutive flotation steps
Air classification	High purity	Low yield
Spontaneous liberation	No extracting chemicals needed	Low yield
Sodium hypochlorite–homogenization	High purity, high yield, suitable for large scale, reduced operation costs	—

Table 3.2 An overview of PHA recovery from various organisms using a halogen-free solvent system.

Microorganism	Initial PHA (%)	Recovery treatment	Purity (%)	Recovery yield (%)	Mw^a before recovery (Da)	Mw after recovery (Da)	Reference
Cupriavidus necator	58.8	Non-halogenated solvent extraction using isoamyl propionate, propyl butyrate, isoamyl valerate or isoamyl isoamylate (isovalerate) (3-methyl-1-butanoate of 3-methyl-1-butanol)	—	90.0	1 000 000	550 000–735 000	57
Cupriavidus necator	60–75	Non-halogenated solvent extraction using isoamyl alcohol	—	98.0	1 000 000	650 000–780 000	58
Cupriavidus necator	—	Supercritical carbon dioxide (1 time), 200 atm, 40 °C, 100 min and 0.2 mL methanol	—	89.0	ND^b	ND	31
Cupriavidus necator	—	Supercritical carbon dioxide (2 times) with 1% (v/v) toluene as a modifier, 200 atm, 30 °C	—	81.0	ND	ND	30
Recombinant *Cupriavidus necator*	—	Non-halogenated solvent extraction using toluene, 1 h, 90 °C	98.0	90.0	1 000 000	390 000	59
Cupriavidus necator	—	Bromelain; pancreatin	89.0	90.0	ND	ND	4
Cupriavidus necator	—	Papain	89.0	78.0	ND	ND	22
Burkholderia sp. PTU9	—	Sonication	92.0	20.0	ND	ND	60
Bacillus flexus	—	Cell fragility by alkaline hydrolysis	—	50.0	ND	ND	61
Bacillus flexus	—	Cell fragility by alkaline hydrolysis	—	94.0	ND	ND	38
Azotobacter vinelandii UWD	77	0.2 M NaOH, 1 h, 30 °C	98.5	92.0	2 200 000	1 900 000	21
Recombinant *Escherichia coli*	32	0.1 M NaOH, 30 °C, 1 h	46.0	80.0	ND	ND	62
Cupriavidus necator	—	0.5 M NaOH, 30 °C, 1 h	75.3	67.2	ND	ND	11
Recombinant *Pseudomonas putida* GPp104	10	0.1 M NaOH, 2 h, 22 °C	44.0	92.0	82 000	82 000	63
Pseudomonas putida	70	1 M NaOH, 3 h, 60 °C	78.0	45.0	ND	ND	64
Cupriavidus necator H16	≈30	0.05 M NaOH, 3 h, 0 rpm, 4 °C	96.6	96.9	169 000	147 000	53
Recombinant *Cupriavidus necator*	≈30	Water, 1 h, 0 rpm, 30 °C	80.6	96.1	169 000	325 000	52
Recombinant *Cupriavidus necator*	≈30	0.05 M NaOH, 1 h, 0 rpm, 4 °C	88.6	96.8	477 000	163 000	65
Comamonas sp. EB172	≈30	Water, 5 h, 0 rpm, 30 °C	48.1	93.5	477 000	854 000	54

^{a}Weight average molecular weight, which was determined by GPC from lyophilized cells.
^{b}Not determined.

impact of process parameters on the effectiveness of PHA extraction procedures have been well proven and studied in the literature, but there is a limitation on concrete data for stated observations about the effect of external factors on PHA recovery.

The extraction method influences the amount of PHA recovered, the simplicity of the subsequent purification steps and eventually the quality of the final product. However, the processing required for effective PHA extraction varies significantly in regards to the PHA-producing organisms as different procedures have various effects on a specific strain and *vice versa*.[3] On the other hand, the PHA content of the produced biomass also plays a dominant role in PHA recovery efficiency because a lower PHA content clearly results in a high recovery cost. This is mainly due to the use of large amounts of digesting agents for PHA extraction which increase the cost of waste disposal.[66]

The final intended application for the PHA will determine the degree of purity of the PHA granules. For example, in medical applications, it is absolutely necessary that the PHA should be free from bacterial endotoxins and other contaminating chemicals and solvents. On the other hand, if the PHA is intended for applications such as mulching film or garbage bags, a lower degree of purity may be acceptable. Regardless of its final applications, the molecular weights of the recovered PHA should be sufficiently high. This is because the thermal processing of the PHA would result in the reduction of its molecular weights to some extent. Therefore, it is important to have PHA resins with as high a molecular weight as possible. In order to obtain PHAs with a high degree of purity, a more stringent recovery process is needed. This often results in PHAs with lower molecular weights. In addition, the recovery yield will be lower. Therefore, the challenge in the recovery process is to maintain the original molecular weights while not compromising the degree of purity for various applications. These criteria have to be achieved in an environmentally friendly manner. Finally, and most importantly, the cost of the recovery process should be economically feasible.[51]

Currently, most PHA extraction processes are based on halogenated solvent extraction which is costly and may cause environmental problems and toxicity to humans. Thus, it seems that a practical commercial extraction system with a clean, simple and efficient process for PHA recovery at a reasonable cost focusing on a non-halogenated solvent extraction-based recovery needs to be developed. However, halogen-free methods require further adjustment, depending on both significant process parameters and external factors influencing their performance, to make the process suitable for polymer recovery on an industrial scale.

References

1. S. Khanna and A. K. Srivastava, *Process Biochem.*, 2005, **40**, 607.
2. R. S. M. S. Karumanchi, S. N. Doddamane, C. Sampangi and P. W. Todd, *Trends Biotechnol.*, 2002, **20**, 72.
3. Y. Chisti and M. Moo-Young, *Enzyme Microb. Technol.*, 1986, **8**, 194.

4. F. M. Kapritchkoff, A. P. Viotti, R. C. P. Alli, M. Zuccolo, J. G. C. Pradella, A. E. Maiorano, E. A. Miranda and A. Bonomi, *J. Biotechnol.*, 2006, **122**, 453.

5. T. R. Hopkins, *Bioprocess Technol.*, 1991, **12**, 57.

6. Y. H. Chen, M. L. Wu and W. M. Fu, *J. Physiol.*, 1998, **507**, 41.

7. S. Y. Lee, J. Choi, K. Han and J. Y. Song, *J. Appl. Environ. Microbiol.*, 1999, **65**, 2762.

8. Y. Poirier, C. Nawrath and C. Somerville, *Biotechnology*, 1995, **13**, 142.

9. D. Byrom, *Polyhydroxyalkanoates*, Hanser, Munich, 1994, pp. 5–33.

10. J. M. Liddell, *Process for the recovery of polyhydroxyalkanoic acid*, United States Patent, 1999, US 5894062.

11. V. Gorenflo, G. Schmack, R. Vogel and A. Steinbuchel, *Biomacromolecules*, 2001, **2**, 45.

12. N. Jacquel, C. W. Lo, Y. H. Wei, H. S. Wu and S. S. Wang, *Biochem. Eng. J.*, 2008, **39**, 15.

13. J. A. Ramsay, E. Berger, B. A. Ramsay and C. Chavarie, *C. Biotechnol. Tech.*, 1990, **4**, 221.

14. E. Berger, B. A. Ramsay, J. A. Ramsay, C. Chavarie and G. Braunegg, *Biotechnol. Tech.*, 1989, **3**, 227.

15. K. S. Roh, S. H. Yeom and Y. J. Yoo, *Biotechnol. Tech.*, 1995, **9**, 709.

16. S. K. Hahn, Y. K. Chang, B. S. Kim and H. N. Chang, *Biotechnol. Bioeng.*, 1994, **44**, 256.

17. J. Choi and S. Y. Lee, *Bioprocess Biosyst. Eng.*, 1997, **17**, 335.

18. Y. Chen, H. Yang, Q. Zhou, J. Chen and G. Gu, *Process Biochem.*, 2011, **36**, 501.

19. J. Yu and L. X. L. Chen, *Biotechnol. Prog.*, 2006, **22**, 547.

20. R. Li, H. Zhang and Q. Qi, *Bioresour. Technol.*, 2007, **98**, 2313.

21. J. Choi and S. Y. Lee, *Biotechnol. Bioeng.*, 1999, **62**, 546.

22. C. H. Lu, *Purification and separation of polyhydroxyalkanoates from bacteria*, Master Thesis, Yuan Ze University, Taiwan, 2006.

23. K. Yasotha, M. K. Aroua, K. B. Ramachandran and I. K. P. Tan, *Biochem. Eng. J.*, 2006, **30**, 260.

24. M. Tamer, M. Moo-Young and Y. Chisti, *Ind. Eng. Chem. Res.*, 1998, **37**, 1807.

25. S. T. Harrison, *Biotechnol. Adv.*, 1991, **9**, 217.

26. F. Donsi, G. Ferrari, E. Lenza and P. Maresca, *Chem. Eng. Sci.*, 2009, **64**, 520.

27. M. Tamer and M. Moo-Young, *Bioprocess Biosyst. Eng.*, 1998, **19**, 459.

28. A. Patist and D. Bates, *Innovative Food Sci. Emerging Technol.*, 2008, **9**, 147.

29. D. A. J. Wase and Y. R. Patel, *J. Chem. Tech. Biotechnol.*, 1985, **35**, 165.

30. K. Khosravi-Darani, E. Vasheghani-Farahani, S. A. Shojaosadati and Y. Yamini, *Biotechnol. Prog.*, 2004, **20**, 1757.

31. P. Hejazi, E. Vasheghani-Farahani and Y. Yamini, *Biotechnol. Prog.*, 2003, **19**, 1519.

32. P. van Hee, A. C. M. R. Elumbaring, R. G. J. M. van der Lans and L. A. M. Van der Wielen, *J. Colloid Interface Sci.*, 2006, **297**, 595.

33. I. Noda, *Process for recovering polyhydroxyalkanoates using air classification, solvent extraction of polyhydroxyalkanoates from biomass facilitated by the use of marginal nonsolvent*, United States Patent, 1998, US 5849854.
34. I. L. Jung, K. H. Phyo, K. C. Kim, H. K. Park and I. G. Kim, *Res. Microbiol.*, 2005, **156**, 865.
35. S. T. Harrison, J. S. Dennis and H. A. Chase, *Bioseparation*, 1991, **2**, 95.
36. Y. Yanagita, N. Ogawa, Y. Ueda, F. Osakada and K. Matsumoto, *Method of collecting highly pure polyhydroxyalkanoate from microbial cells*, United States Patent, 2008, PCT/JP2004/000416.
37. K. Narasimhan, A. C. Cearley, M. S. Gibson and S. J. Welling, *Process for the solvent-based extraction of polyhydroxyalkanoates from biomass*, United States Patent, 2008, US 7378266.
38. W. J. Page and A. Cornish, *Appl. Environ. Microbiol.*, 1993, **59**, 4236.
39. S. P. Ouyang, R. C. Luo, S. S. Chen, Q. Liu, A. Chung, Q. Wu and H. Q. Chen, *Biomacromolecules*, 2007, **8**, 2504.
40. H. Wang, X. Zhou, Q. Liu and G. Q. Chen, *Appl. Microbiol. Biotechnol.*, 2011, **89**, 1497.
41. J. Han, M. Li, J. Hou, L. Wu, J. Zhou and H. Xiang, *Saline Syst.*, 2010, **6**, 1.
42. K. H. Chia, T. F. Ooi, A. Saika, T. Tsuge and K. Sudesh, *Polym. Degrad. Stab.*, 2010, **95**, 2226.
43. J. Xu, B. H. Guo, R. Yang, Q. Wu, G. Q. Chen and Z. M. Zhang, *Polymer*, 2002, **43**, 6893.
44. G. Q. Chen, *Chem. Soc. Rev.*, 2009, **38**, 2434.
45. L. Luo, X. Wei and G. Q. Chen, *J. Biomater. Sci.*, 2009, **20**, 1537.
46. M. N. Belgacem and A. Gandini, *Monomers, polymers and composites from renewable resources*, Elsevier, Amsterdam, 2008, pp. 39–66.
47. Y. Kathiraser, M. K. Aroua, K. B. Ramachandran and I. K. P. Tan, *J. Chem. Technol. Biotechnol.*, 2007, **82**, 847.
48. R. A. J. Verlinden, D. J. Hill, M. A. Kenward, C. D. Williams and I. Radecka, *J. Appl. Microbiol.*, 2007, **102**, 1437.
49. S. P. Valappil, S. K. Misra, A. R. Boccaccini, T. Keshavarz, C. Bucke and I. Roy, *J. Biotechnol.*, 2007, **132**, 251.
50. O. Wolf, M. Crank, M. Patel, F. Marscheider-Weidemann, J. Schleich, B. Husing and G. Angerer, *Techno-economic feasibility of large-scale production of bio-based polymers in Europe*, Institute for Prospective Technological Studieso, European Communities: Spain, 2005.
51. B. Kunasundari and K. Sudesh, *eXPRESS Polym. Lett.*, 2011, **5**, 620.
52. M. Mohammadi, M. A. Hassan, L. Y. Phang, H. Ariffin, Y. Shirai and Y. Ando, *Biotechnol. Lett.*, 2012, **34**, 253.
53. M. Mohammadi, M. A. Hassan, L. Y. Phang, Y. Shirai, C. M. Hasfalina, H. Ariffin, A. A. A. Amirul and A. Syairah, *Environ. Eng. Sci.*, 2012, **29**, 783.
54. M. Mohammadi, M. A. Hassan, L. Y. Phang, Y. Shirai, C. M. Hasfalina and H. Ariffin, J, *Cleaner Prod.*, 2012, **37**, 353.
55. M. R. Zakaria, H. Ariffin, N. A. Mohd Johar, A. A. Suraini, H. Nishida, Y. Shirai and M. A. Hassan, *Polym. Degrad. Stab.*, 2010, **95**, 1382.

56. M. S. Ghatnekar, J. S. Pai and M. Ganesh, *J. Chem. Technol. Biotechnol.*, 2002, **77**, 444.

57. P. E. Mantelatto and N. A. Durao, *Process for extracting and recovering polyhydroxyalkanoates (Phas) from cellular biomass*, 2008, US20080193987A1.

58. P. E. Mantelatto, A. M. Duzzi, T. Sato, N. A. Durao, R. V. Nonato, C. Rocchiccioli and S. M. Kesserlingh, *Process for recovering polyhydroxialkanoates ('Phas') from cellular biomass*, 2005, WO/2005/052175A2.

59. K. Kinoshita, Y. Yanagida, F. Osakada, Y. Ueda, K. Narasimhan, A. C. Cearley, K. Yee and I. Noda, *Process for producing polyhydroxyalkanoate*, European Patent Application, 2007, EP1739182.

60. M. S. Divyashree, T. R. Shamala and N. K. Rastogi, *Biotechnol. Bioprocess Eng.*, 2009, **14**, 482.

61. M. S. Divyashree and T. R. Shamala, *Indian J. Microbiol.*, 2010, **50**, 63.

62. P. T. Voon, *Environmental friendly alternative methods for the recovery of intracellular polyhydroxyalkanoates (PHA)*, Master Thesis, Universiti Putra Malaysia, Malaysia, 2005.

63. X. Jiang, J. A. Ramsay and B. A. Ramsay, *J. Microbiol. Methods*, 2006, **67**, 212.

64. Y. H. Yang, C. Brigham, L. Willis, C. Rha and A. Sinskey, *Biotechnol. Lett.*, 2011, **33**, 937.

65. M. Mohammadi, M. A. Hassan, Y. Shirai, C. M. Hasfalina, H. Ariffin, Y. Lian-Ngit, T. Mumtaz, M. L. Chong and L. Y. Phang, *Sep. Sci. Technol.*, 2011, **47**, 1.

66. H. Salehizadeh and M. C. M. Van Loosdrecht, *Biotechnol. Adv.*, 2004, **22**, 261.

CHAPTER 4

Blends of Polyhydroxyalkanoates (PHAs)

HEMA RAMACHANDRAN,[a,b] SHANTINI KANNUSAMY,[a] KAI-HEE HUONG,[a,d] RENNUKKA MATHAVA[a] AND A.-A. AMIRUL*[a,c,d]

[a] School of Biological Sciences, Universiti Sains Malaysia, 11800 Penang, Malaysia; [b] Quest International University Perak, 30250 Ipoh, Perak; [c] Centre for Chemical Biology, Universiti Sains Malaysia, 11800 Penang, Malaysia; [d] Malaysian Institute of Pharmaceuticals and Nutraceuticals, MOSTI, Malaysia
*Email: amirul@usm.my

4.1 Introduction

The current prominence of sustainability, eco-efficiency and green chemistry has generated a tremendous search for materials that are renewable and environmentally friendly. Biodegradable polymers offer a sustainable alternative to petroleum-derived plastics. These polymers largely consist of ester, amide and other functional groups. Biodegradable polymers can be classified into four categories according to their synthesis process: (i) natural polymers (*e.g.*, starch, cellulose, chitosan and protein), (ii) polymers synthesized from natural monomers such as polylactide acid (PLA), (iii) polymers conventionally and chemically synthesized from monomers obtained from petrochemical products (*e.g.*, polycaprolactone) and (iv) polymers synthesized by microbes such as polyhydroxyalkanoates (PHAs) (Figure 4.1). The performance of these biodegradable polymers is usually enhanced and altered *via* blending.[1,2]

RSC Green Chemistry No. 30
Polyhydroxyalkanoate (PHA) Based Blends, Composites and Nanocomposites
Edited by Ipsita Roy and Visakh P M
© The Royal Society of Chemistry 2015
Published by the Royal Society of Chemistry, www.rsc.org

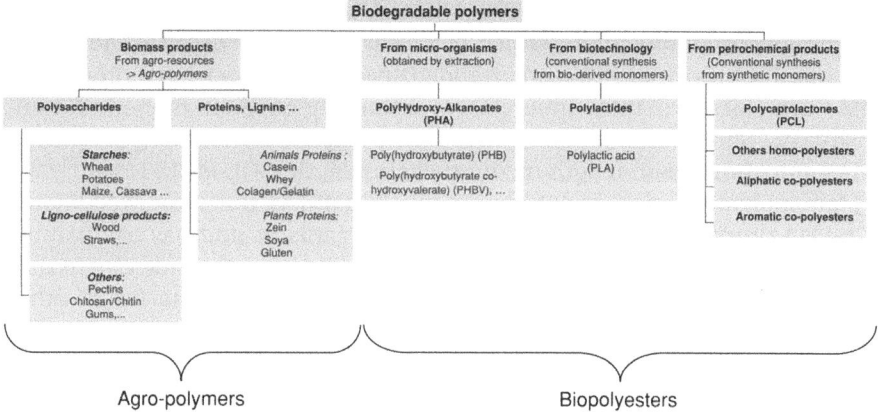

Figure 4.1 Classification of the main biodegradable polymers.[1]

Polymer blending has attracted the attention of researchers because polymers with extraordinary properties obtained by chemical synthesis are more expensive than existing polymers and blending operations. Furthermore, a wise choice and combination of the polymeric materials in specific amounts may lead to the fabrication of blend materials with desirable properties. There are various numbers of polymers that can be combined to form blends with different physical properties. The characteristics of the polymeric blend are influenced by the nature of the dispersed and dispersion phases, the volume ratio of the phases, the sizes and size distributions of the particles of the dispersed phase and interfacial adhesion.[3] One of the popular questions being addressed regarding the polymer blend is the miscibility between the components. The blends formed can be miscible, partially miscible or fully immiscible. The miscible polymer blend is formed by choosing polymers with compatible chemical structures which are capable of specific interactions.[4]

Polyhydroxyalkanoates (PHAs) are members of a greatly fascinating family of microbial polyesters of 3-, 4-, 5- and 6-hydroxyacids that show potential for various industrial and medical applications due to their wide range of characteristics.[5,6] These polymers with imperative ecosystem roles and high biotechnological potential are synthesized naturally by a diverse range of bacterial species from at least 75 different genera.[7,8] The polymers are usually accumulated as insoluble inclusions in the cytoplasm of bacterial cells during the depletion of essential nutrients such as nitrogen, magnesium or phosphorus in the presence of abundant carbon sources.[9]

Although PHAs are considered to be consumer-oriented and environmentally friendly biopolymers due to their biodegradability and biocompatibility, commercialisation of these biopolymers is stringently dependent on the material properties that satisfy the requirements of the targeted market application. At present, more than 200 different monomer constituents are found either as homopolyesters or in combination as copolyesters.[10]

The wide substrate range of PHA synthase has resulted in the versatility of the monomer compositions, which is a clear advantage because the monomer variation provides PHAs an extended spectrum of associated properties. PHAs are classified into three classes according to their monomer compositions: Short-chain-length (SCL)-PHAs, medium-chain-length (MCL)-PHAs and short-chain-length-medium-chain-length (SCL-MCL)-PHAs.[11,12]

Homopolymer poly(3-hydroxybutyrate) [P(3HB)] is the most well-known SCL-PHA produced by wide ranges of microorganisms and has comparable material properties with polypropylene.[13] Manipulating the side chains and compositions of the P(3HB) homopolymer through incorporation of other monomers can generate different types of polymers with favourable material properties as the polymers will confer less stiffness and tougher properties. Among the SCL-PHAs that have been studied with such material properties are poly(3-hydroxybutyrate-*co*-3-hydroxyvalerate) [P(3HB-*co*-3HV)] and poly(3-hydroxybutyrate-*co*-4-hydroxybutyrate) [P(3HB-*co*-4HB)] copolymers.[14] MCL-PHAs are more elastomer in nature compared to SCL-PHAs, which are often stiff and brittle. Incorporating both monomers will result in SCL-MCL-PHA copolymers exhibiting properties between the two states which will depend on the different proportions of SCL and MCL monomers.[15] Poly(3-hydroxybutyrate-*co*-3-hydroxyhexanoate) [P(3HB-*co*-3HHx)] copolymer is one of the successful SCL-MCL-PHAs that is produced on an industrial scale.[12]

Various PHA blends have been developed to improve the performance and to offset the high price of PHAs. The blending of PHAs will offer more scope to expand their range of applications. In this chapter, some features of blends of SCL-PHAs with PLA, starch, PCL, cellulose and chitosan will be discussed.

4.2   Production of Blends of PHAs

4.2.1   PHAs/PLA Blends

The P(3HB)/PLA blend is one of the most studied blends, which exhibits mechanical properties that are intermediate between the individual components. Although PLA and P(3HB) are biodegradable polymers synthesized from renewable resources, their potential applications are hampered due to their brittleness and the formation of very large spherulites.[16,17] P(3HB)/PLA blends were studied as early as 1996 to explore their miscibility, crystallisation, morphology, mechanical properties and biodegradation behaviour. P(3HB)/PLA blends with different compositions (100/0, 80/20, 60/40, 40/60, 20/80 and 0/100, wt%) were prepared by casting a film from a common solvent, chloroform, at room temperature.[18]

A melt-crystallisation method had been employed by Ohkoshi and coworkers[19] to prepare blends of atacticP(3HB) with PLA. The atacticP(3HB) [ataP(3HB)] is synthesized by the polymerisation of racemic β-butyrolactone and is a completely amorphous polymer. The solvent-casted film of ataP(3HB)/PLA blends were inserted between two Teflon sheets (0.2 mm) and compression-molded on a Mini Test Press by heating at 200 °C for 1 minute under a pressure of 75 kg cm^{-2}. After melting, samples were kept at a given crystallisation temperature (T_c) and isothermally crystallised for 3 days. The melt-crystallised films were about 0.2 mm in thickness and stored at 220 °C.

The effect of adding PLA on the toughness of P(3HB) had also been studied by preparing the blends using a melt-compounding method. The P(3HB)/PLA blends at different concentrations (100/0, 75/25, 50/50, 25/75, and 0/100, wt%) were mixed in a Haake polylab torque rheometer at 175 °C for 10 minutes at a constant rotor speed of 50 rpm. The materials were hot-pressed at 180 °C using a 20 ton lab press and then quenched to room temperature to obtain 1 mm and 3 mm sheets.[20] A plasticized P(3HB)/PLA: 75/25 wt% blend was also prepared through the melt-compounding method to create a new type of eco-friendly blend material suitable for single-use applications, such as fast food packaging. Polyester plasticizer (Lapol 108) was added at two different concentrations (5 and 7 wt%).[21]

It had been reported that the commercial viability of pure P(3HB-*co*-3HV) copolymer and PLA alone is difficult, therefore blends of these polymers were prepared to get a material with balanced toughness properties.[22] Another study reported that the blend of P(3HB-*co*-3HV)/PLA provides a way of improving and tailoring the foam morphologies without compromising the biodegradability.[23] P(3HB-*co*-3HV)/PLA blends were prepared by melt mixing in a mixer. All materials were dried in a vacuum prior to use. The resulting materials were then ground into pellets and then were sent for compression molding. The compression molding was carried out under a hydraulic pressure.[24] Meanwhile, Zhao and co-workers[23] reported the preparation of a P(3HB-*co*-3HV)/PLA blend using a co-rotating twin-screw extruder. The melt mixing was carried out above the glass transition temperature (T_g) of amorphous polymer. The twin-screw extruder was used to ensure that all the specimens undergo the same thermal-mechanical history. Nanda and co-workers[22] reported the production of a P(3HB-*co*-3HV)/PLA blend using a micro compounder. The polymers were dried in a vacuum prior to use. In order to obtain the desired samples, the molten mass of the P(3HB-*co*-3HV)/PLA blends were transferred to a mini injection moulder.

Since PLA exhibited properties such as brittleness, a slow rate of crystallisation and enzymatic hydrolysis, the practical application of PLA was limited. P(3HB-*co*-4HB) copolymer seems to be a suitable candidate as this copolymer has relatively good toughness, flexibility and high biodegradation rate. PLA was blended with P(3HB-*co*-4HB) copolymer by melt compounding. This had resulted in a more favorable blending method as the modified properties were expected to overcome the drawbacks and provide feasible applications.[25]

4.2.2 PHAs/PCL Blends

PCL is semi-crystalline polyester that exhibits good mechanical properties, biodegradability, biocompatibility and miscibility with various polymers.

Processability of the blend can also be enhanced as PCL lowers the elastic modulus by acting as a polymeric plasticizer. However, it is ductile and has a lower melting point, T_m, of 55 °C. Blending of PCL and P(3HB) offers a good option to improve the performance of both homopolymers.[26,27] Antunes and Felisberti[26] prepared a blend of P(3HB)/PCL by melting the mixture in an internal mixer with compositions of PCL varying from 0 to 30 wt% to study the miscibility, morphology and physical–chemical properties of these systems. The solvent-casting method is a common method to prepare blends of polymers. This method had been employed by Chee *et al.*[28] and Lovera *et al.*[27] to prepare a P(3HB)/PCL blend whereby chloroform was used as a solvent to dissolve the polymers. Lovera and co-workers[27] blended high molecular weight P(3HB)/PCL and P(3HB)/low molecular weight chemically modified PCL (mPCL) by dissolving them in chloroform at 40 °C to study the morphology, crystallisation, and enzymatic degradation of the blends. P(3HB-*co*-3HV)/PCL polymer blends were also prepared by mixing in chloroform and stirring overnight. The blends were then casted on glass plates and were left to dry under vacuum.[29] The same method was reported by Wessler *et al.*[30] and Qiu *et al.*[31] to prepare the P(3HB-*co*-3HV)/PCL blends. Although the solvent-casting method could avoid thermal degradation of the polymers, due to the presence of residual chloroform, it was found not suitable for biomedical applications.[32]

An attractive route to prepare P(3HB)/PCL through injection molding was employed to obtain a material with high flexibility and good biodegradability to be applied in packaging. Dogbone-shaped specimens of P(3HB)/PCL: 70/ 30 wt% blends were obtained in a Battenfeld 250 Plus injection molding machine with a clamping force of 25 tons, plasticizing capacity of 9.5 g s^{-1}, and $L/D = 16$. The screw speed was maintained at 200 rpm with the temperature raised from 135 °C in the first barrel zone to 165 °C in the die zone.[33]

Jenkins and co-workers[32] reported the usage of supercritical carbon dioxide as a solvent in the melt blending process of P(3HB-*co*-3HV) copolymer and PCL. This method did not result in solvent residue and minimised the possibility of thermal degradation. In this method, pellets of PCL and P(3HB-*co*-3HV) polymers were introduced in a sealed vessel with a mechanical stirrer. Initially, the vessel was heated to allow the PCL to melt while P(3HB-*co*-3HV) copolymer remained as a semi-crystalline solid. The vessel was then pressurised with CO_2 where both the polymers melted. Then, the blends were stirred to ensure mixing and to allow the dissolution of CO_2 in the polymer. The same author also reported the preparation of P(3HB-*co*-3HV)/PCL using mechanical blends. In this method, the pellet forms of PCL and P(3HB-*co*-3HV) copolymers were used. The blend was produced using a two-roll mill. The rollers were pre-heated followed by the addition of PCL and later P(3HB-*co*-3HV) copolymer. This allowed the PCL to melt first and limit the thermal degradation of the P(3HB-*co*-3HV) copolymer. The polymers were then mixed together.

4.2.3 PHAs/Cellulose Derivatives

Cellulose is a linear polymer made up of glucose molecules linked by β-(1,4)-glycosidic linkages. Due to its biocompatibility, hydrophilicity and chirality, cellulose has been widely used in the immobilisation of proteins and antibodies for the separation of enantiomeric molecules and formation of cellulose composites with synthetic polymers and biopolymers.[34] Cellulose derivatives also have attracted much interest for their compatibility with P(3HB).[2] Ethyl cellulose (EtC) is also a biomaterial like P(3HB) that is approved by the FDA (Food and Drug Administration) and is widely used as a blood coagulant, in coatings for pharmaceutical tablets and matrices for poorly soluble drugs. Zhang and co-workers[35] investigated the miscibility, thermal behaviour and morphological structure of P(3HB) with ethyl cellulose (EtC) blends. Thin films of P(3HB)/EtC blends (100/0, 80/20, 60/40, 40/60, 20/80 and 0/100) were prepared by casting from 3% (wt/v) solutions of the two components in chloroform. According to Zhang and Deng,[36] blending of P(3HB) with hydroxyethyl cellulose acetate (HECA) would create a new biodegradable polymer. The blends were prepared as reported by Zhang *et al.*[35]

Phase behaviour and crystallisation kinetics for the binary blend P(3HB)/cellulose propionate (CP) were performed by Maekawa *et al.*[37] Cellulose acetate butyrate (CAB), which has a combination of high T_m (160 °C) and T_g (113 °C), is an important thermoplastic cellulose ester that can biodegrade in a natural environment. In an attempt to make the best use of degradable polyester P(3HB), Wang *et al.*[38] blended P(3HB) with CAB and studied the relationship between the blend morphology and its physical properties.

A cold-drawing procedure was applied to P(3HB)-based blends with cellulose propionate (CP) to obtain high performance blend films. The P(3HB)-rich film blends containing 0–30 wt% CP were cut into 3×1 cm strips, hot-pressed between two Teflon sheets at 200 °C and then rapidly quenched in ice-water (4 °C to 10 °C) to preserve the state of amorphous preform. The melt-quenched strip was attached to a portable drawing device designed to stretch to a maximum of 30 times by hand operation. In contrast, CP-rich film blends (50–100 wt% CP) were immediately placed into a heating oven adjusted to the given drawing temperature (60 °C to 175 °C) after attachment to the drawing device. Then, they were kept for 1 minute to be thermally stabilised and then stretched to a given draw ratio.[39]

The characteristics and bending performance of the novel electroactive polymer blend P(3HB)/cellulose in terms of free bending displacement output, electrical power consumption and lifetime were studied by Zhijiang *et al.*[34] Cellulose is a potential smart material which responds to an *ith* electric field. Electro-active polymer (EAP) was prepared by dissolving cellulose and P(3HB) in trifluoroacetic acid. The solution was cast to form a film followed by the deposition of a thin gold electrode on both sides of the film.

A blend of P(3HB-*co*-3HV) copolymer and cellulose was prepared using an ionic solvent by Hameed and co-workers.[40] In their study, the micro granular cellulose and P(3HB-*co*-3HV) copolymer were dried under vacuum in order to remove the moisture. Cellulose was dissolved using 1-butyl-3-methylimidazolium chloride ([BMIM]CI). The mixture was then heated and stirred to obtain a homogeneous solution. The P(3HB-*co*-3HV) copolymer was then added and coagulated in water. The mixture was then stirred at 100 °C. In order to obtain the film, the viscous solution of cellulose and P(3HB-*co*-3HV) copolymer was casted in between two glass plates and then soaked in a water bath to allow the ionic solution to evaporate.

Gilmore *et al.*[41] reported the preparation of a blend of cellulose acetate esters and P(3HB-*co*-3HV) copolymer using a solvent-casting method. P(3HB-*co*-3HV) copolymer and cellulose acetate esters were dissolved separately in chloroform and then dried under vacuum. Lotti and Scandola[42] reported the preparation of a blend of cellulose acetate butyrate and P(3HB-*co*-3HV) copolymer whereby the components were mixed in an injection molding machine. The temperature was kept low in order to minimise the thermal degradation of P(3HB-*co*-3HV) copolymer.

4.2.4 PHAs/Starch Derivatives

Starch is a renewable resource that comprises a linear polymer (amylose) and a branched polymer (amylopectin). It is an attractive biopolymer due to its low cost, low density, non-abrasive nature and biodegradability. It is readily available from agricultural sources such as maize, potatoes, tapioca, rice and wheat. Starch is widely used in food and non-food applications. Starch is therefore classified as inexpensive and a commodity material. The hydrophilic nature of starch limits the development of starch-containing materials; therefore, starch is usually blended with other polymers. Blending the starch with biodegradable and more hydrophobic polymers produces blends that are completely biodegradable under biologically active environments.[43] The main reason for adding starch to PHAs is to reduce the overall cost besides improving its properties. Since PHAs are hydrophobic and starch is hydrophilic, there will be a poor adhesion between these two polymers thus contributing to poor mechanical properties.[44]

P(3HB)/starch blend was prepared either by a conventional solvent-casting method or by melt processing methods, such as injection molding and compression molding after compounding. Two types of maize starch, Starch 1 (containing 70% amylose) and Starch 2 (containing 72% amylopectin) were blended with P(3HB) using a melt compounding method at a ratio of 70/30 wt% and they were characterised in terms of their morphology, structure, thermal, rheological and mechanical properties.[45] The thermal behaviour and morphological structure of solvent-cast films of P(3HB)/starch acetate (SA) blend with weight ratios of 100/0, 80/20, 60/40, 40/60, 20/80 and 0/100 had been studied by Zhang *et al.*[46] Ismail and Gamal[47] prepared P(3HB)/starch composites with different starch contents to investigate their mechanical and swelling in water properties and to evaluate the capability of Actinomycetes to degrade this polymer under different cultural conditions. Stearic acid was added as a compatibilizer and glycerol as a plasticizer to improve the adhesion of filler (starch) to polymer matrix (P(3HB).

Ramsay and co-workers[48] reported the blending of P(3HB-*co*-19%3HV) copolymer with starch. The copolymer was blended at 20 rpm prior to mixing with starch. The mixture was then hot-pressed at high temperature and high pressure to form sheets. Magalhaes and Andrade[49] reported the preparation of P(3HB-*co*-3%3HV) copolymer with starch whereby the starch was mixed with glycerol before it was homogenized using a blender. Then, the blender was used to mix the starch with copolymer in appropriate ratios. The processing was carried out in a co-rotating twin-screw extruder using a temperature profile. Later, the materials were compression-molded by heating and cooling. Koenig and Huang[50] reported the preparation of starch derivative with P(3HB-*co*-3HV) copolymer. Starch acetate derivatives were prepared by reaction of destructured starch with acetic anhydride in acetic acid. Dimethylsulfoxide (DMSO) was added to break up the granular structure. The polymer and starch blends were made by solution blending with methylene chloride.

In another study by Seves and co-workers,[51] P(3HB-*co*-3HV) copolymer was blended with starch valerate. Starch valerate was produced through an esterification reaction under a nitrogen stream. The esterification of starch was confirmed using IR analysis. This material was used in an amorphous form in blend preparation as a diluent of the semi-crystalline biodegradable P(3HB-*co*-3HV) copolymer. A series of binary blends with certain weight ratios were prepared in the form of sheets by mixing the molten polymers in a Gimac single-screw microextruder. Preparation of corn starch and P(3HB-*co*-3HV) copolymer blend films was reported by Reis and co-workers.[52] The blend films were prepared using a conventional solution casting technique.

Zhang *et al.*[53] aimed to overcome the lower crystallisation rate and large crystal size of the P(3HB-*co*-4HB) copolymer by blending the P(3HB-*co*-4HB) copolymer with corn starch. Besides that, the high production cost of P(3HB-*co*-4HB) had also led to the limitation of the applications. Therefore, corn starch, which is a polysaccharide found in nature, was selected as a raw material in the blending. The blends were fabricated by melt mixing and injection molding.

4.2.5 PHAs/Chitosan

Chitosan (CS) is a natural biocompatible cationic polysaccharide that has common comonomer units, 2-deoxy-2-acetamido-D-glucose and 2-deoxy-2-amino-D-glucose. These units are linked by β-1,4-bonding. Chitosan exhibits mild antimicrobial activity arising from its cationic residue, which is an important characteristic for its application as a biomaterial. It can suppress the metabolism of bacteria when it adheres to the bacterial cell wall. Chitosan has also attracted the attention of researchers due to its biodegradability *in vivo* and biocompatibility. A chemically modified

P(3HB)/chitosan blend can be prepared because chitosan has functional groups, such as hydroxyl, amine and amide groups.[54,55]

Ikejima and co-workers[54] prepared P(3HB)/chitosan blend films in order to investigate the effect of deacetylation on the crystallisation behaviour of P(3HB). Chitosan is a copolysaccharide with a high degree of deacetylation. The solvent-casting method was employed to prepare the P(3HB)/chitosan blend films. P(3HB) and chitosan were dissolved separately in HFIP (1,1,1,3,3,3-hexafluoro-2-propanol) before blending. After the chitosan solutions were well homogenized, they were mixed with the P(3HB) solution. Cheung *et al.*[56] also reported that chitosan and P(3HB-*co*-3HV) copolymer are miscible due to the presence of functional groups such as hydroxyls, amines and amides that could be modified in the chitosan. The preparation of this P(3HB-*co*-3HV)/chitosan blend involved a drying method. In this method, both the polymer and chitosan were also dissolved in HFIP. After the chitosan solution had dissolved completely, the P(3HB-*co*-3HV) copolymer solution was added. A thin film of this blend was prepared by casting from HFIP with acetic acid by slow evaporation. The blends were then dried under vacuum.

Hu *et al.*[55] presented the first report on the preparation and characterisation of P(3HB) grafted with chitosan *via* ozone treatment. The P(3HB) membrane was treated with ozone to graft acrylic acid, followed by the esterification of chitosan. The process was started by placing a piece of membrane in an Erlenmeyer flask, which was then flushed with air containing 10.2 g m^{-3} of ozone for a specified time. Unreacted ozone was removed by evacuating the samples for 2 minutes and then soaking them at 65 °C in an aqueous solution containing either 5 or 10% acetic acid, 0.2 M H$_2$SO$_4$ (sulfuric acid) and 1 mM FeSO$_4$ (ferum sulfate). After a specified time, the samples were retrieved and rinsed with adequate double-distilled water three times, followed by soaking in double-distilled water at 75 °C for 24 hours. In the first 10 hours, the water was replenished every 2 hours; after 10 hours, the water was replenished every 4 hours. These procedures were carried out to remove unreacted AA (acetic acid) and the homopolymer of AA (acetic acid).

The properties of P(3HB-*co*-4HB) copolymer can be altered from hard crystalline to elastic rubber by varying the 4HB monomer composition. The crystallinities and tensile strength of the P(3HB-*co*-4HB) copolymer decreased with an increase of 4HB monomer from 0 to 27% whereas the elongation at break increased from 5% to 444%. However, increases in crystallinity and tensile strength were observed for polymers with 4HB compositions from 64–100%.[57] Since P(3HB-*co*-4HB) copolymer is hydrophobic in nature and is absent of bioactivity, adding chitosan, which is hydrophilic and antimicrobial, into P(3HB-*co*-4HB) serves the purpose of the blending. This is significant as water adsorption and antimicrobial activity is important especially in wound healing applications. The blend material of P(3HB-*co*-4HB) with chitosan was fabricated using a simple solvent-casting method.[58]

4.3   Characterisation of Blends of PHAs

4.3.1   PHAs/PLA Blends

Zhang *et al.*[18] studied the miscibility, crystallisation, mechanical properties and hydrolytic degradation of P(3HB)/PLA blends. The results showed that the blends prepared through the solvent-casting method were immiscible in the amorphous state over the range of compositions studied. However, blends prepared through the melt-compounding method showed some evidence of greater miscibility, which was assumed to be due to the trans-esterification of P(3HB) and PLA at high temperatures that had resulted in the *in situ* formation of a P(3HB)/PLA block. Addition of PLA was found to have a greater effect on the crystallisation of P(3HB), especially when its content was relatively high. The growth of P(3HB) spherulites decreased with the addition of amorphous PLA and the blend of P(3HB)/ PLA: 20/80 wt% did not crystallise at all. The mechanical properties of P(3HB)/PLA were improved as indicated by a reduction in modulus and stress at break and an enhancement of the elongation at break. Faster hydrolytic degradation could be observed as the P(3HB)/PLA blend films were more hydrophilic than the P(3HB) films and the PLA component was quickly degraded.

According to Koyama and Doi,[59] the miscibility of the P(3HB)/P[(S)-LA] blend was strongly influenced by the molecular weight of the P[(S)-LA] component. Two phases in the melt at 200 °C were observed for the P(3HB)/ P[(S)-LA] blend with P[(S)-LA] of molecular weight values above 20 000. The P(3HB)/P[(S)-LA] blend with the P[(S)-LA] of molecular weight values below 18 000 exhibited greater miscibility. However, the T_m data did not correlate with the miscibility of the blend studied. Ohkoshi *et al.*[19] reported that the X-ray crystallinities of the PLA components in the melt-crystallised P(3HB)/ PLA blend films increased with the addition of a small amount of ataP(3HB). This was a contradictory finding as in most cases, the crystal growth rate of crystallisable component was depressed by mixing amorphous components. It could be suggested that the addition of the completely amorphous ataP(3HB) component had accelerated the growth rate of PLA spherulites by increasing the chain mobility of the crystallisable PLA component.

Vogel *et al.*[60] studied the phase separation in blends of P(3HB)/PLA as a function of the blend composition by Fourier transform infrared (FT-IR) imaging spectroscopy (Figure 4.2), which illustrated that the P(3HB)/PLA: 50/ 50 wt% blend was phase separated while the P(3HB/PLA): 30/70 wt% blend was a compatible one-phase system. The elongation at break of P(3HB)/PLA: 100/0 wt% and P(3HB)/PLA: 0/100 wt% was 2% and 150%, respectively. The blend of P(3HB)/PLA: 20/80 wt% exhibited an elongation at break of 125% because it was dominated by a PLA with much better mechanical deformation properties.

It was reported by Nanda and co-workers[22] that the addition of PLA to P(3HB-*co*-3HV) copolymer caused the tensile strength and modulus of the blend to gradually increase. An increase in tensile strength and

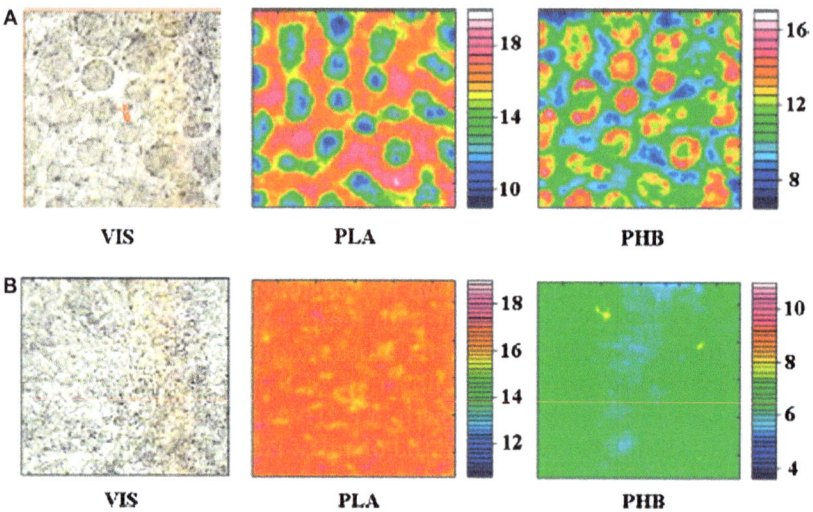

Figure 4.2 (A) Visual image (left), PLA-specific FT-IR image (center), and P(3HB)-specific FT-IR image (right) of a P(3HB)/PLA: 50/50 wt% blend. (B) Visual image (left), PLA-specific FT-IR image (center), and P(3HB)-specific FT-IR image (right) of a P(3HB)/PLA: 30/70 wt% blend.[59]

modulus might be because of the high strength and modulus of the PLA phase. It was also reported that there was a good transfer of stress from one phase to another in the blend. However, Zhao *et al.*[23] reported that increasing the P(3HB-*co*-3HV) copolymer content from 0 to 45% reduced the tensile strength of P(3HB-*co*-3HV)/PLA blends from 54.99 to 42.23 MPa. It was also reported that this might be because of the P(3HB-*co*-3HV) copolymer, which had relatively low tensile strength and relatively weak polymer interfaces.

Noda *et al.*[61] reported that the addition of 10% PHA to PLA increased the elongation of the blend due to the presence of the amorphous phase of the blend. The same trend was reported by Nanda *et al.*[22] whereby the elongation of all P(3HB-*co*-3HV)/PLA blends were more than the neat polymers. Maximum elongation was observed for the P(3HB-*co*-3HV)/PLA: 60/40 wt% blend, where the elongation was reported to be 148 and 250% more than the neat P(3HB-*co*-3HV) copolymer and PLA, respectively. Zhao *et al.*[23] also reported that the strain at break increased as the P(3HB-*co*-3HV) copolymer was incorporated into the PLA. The impact strength of neat PLA was recorded as 30 J m^{-1} and for neat P(3HB-*co*-3HV), it was 49 J m^{-1}. Although the elongation at break reported for the P(3HB-*co*-3HV)/PLA blend was higher than the individual polymer, the impact strength, however, was nearly equal to that of the PLA.[22] Zembouai *et al.*[24] reported that the Young's modulus decreased as the amount of PLA increased. It was also reported that for the blends P(3HB-*co*-3HV)/PLA: 75/25 wt% and 50/50 wt%, the elongation at break was lower than for the neat PLA. However, for the blend

(P(3HB-*co*-3HV)/PLA: 25/75 wt%, it showed an increase in elongation at break compared to the neat PLA (Table 4.1). It was reported that the slight improvement might be because of the P(3HB-*co*-3HV) particles acting as a filler to enhance PLA properties.

Thermal gravimetric analysis (TGA) is known as a popular method to determine the thermal stability of certain polymers. It was reported by Neill and Leiper[62] that the thermal degradation of P(3HB-*co*-3HV) copolymer occurred through a random chain scission at ester groups whereas the thermal degradation of PLA occurred through the cleavage of a bond on the backbone to form a cyclic oligomer. Zembouai *et al.*[24] reported that the degradation of each polymer occurred separately and it was found that the PLA was more thermally stable than the P(3HB-*co*-3HV) copolymer. They also reported that the decomposition temperature, T_d, for all the blends was between the PLA and P(3HB-*co*-3HV) copolymer. Therefore, the thermal stability of the P(3HB-*co*-3HV)/PLA blends could be improved by increasing the amount of PLA. The same trend was reported by Nanda *et al.*[22] It was reported that the TGA curve for pure P(3HB-*co*-3HV) copolymer and PLA was 290 °C and 375 °C, respectively. For the P(3HB-*co*-3HV)/PLA blend, two weight loss peaks were observed, one at 290 °C and the other at 375 °C representing the P(3HB-*co*-3HV) and PLA phase, respectively, indicating that P(3HB-*co*-3HV) copolymer and PLA tend to degrade individually.

In a study by Nanda and co-workers,[22] the thermal properties of the blend were evaluated using differential scanning calorimetry (DSC). The T_m of P(3HB-*co*-3HV) copolymer was reported as 152 °C and for PLA, it was reported as 170 °C. The T_m for the blend was found to be in between these two polymers. Miscibility of any polymer can be determined by evaluating the T_g. A single T_g indicates the miscibility of the polymer. One T_g that corresponded to PLA was reported and the T_g for P(3HB-*co*-3HV) copolymer was poorly observed. They also reported that when the content of P(3HB-*co*-3HV) copolymer was increased, reduction in T_g from 60 °C to 45 °C was observed. The same observation was reported by Richards *et al.*[63] and Modi *et al.*[64]

It was reported by Nanda *et al.*[22] that the degree of crystallinity of any polymer blends depends on the orientation, the flexibility of the polymer chain and nucleation. They also reported a decrease in the T_c of PLA from 119 °C to 83 °C with an increase of P(3HB-*co*-3HV) copolymer content. However, no increase in the degree of crystallinity of PLA was observed. The presence of PLA prevented the crystal growth of P(3HB-*co*-3HV) copolymer in the blend. In order to determine the transparency of any material, UV spectroscopy was used. It was reported that the lower the transparency of any material, the greater the absorbance of radiation and the lower the transmittance. PLA, being transparent, has low absorbance. As the content of HV increased in the blend, the transparency decreased. This showed that P(3HB-*co*-3HV) copolymer has a dense structure that does not allow light to pass through it.

Table 4.1 Thermal and mechanical properties of various PHA blends prepared using the melt compounding method.

PHA blends	Mechanical properties			Thermal properties		
	Tensile strength (MPa)	Elongation at break (%)	Young's modulus (MPa)	T_g^f (°C)	T_c^g (°C)	T_m^h (°C)
P(3HB)/PLA[a]						
100/0	31	7.3	1950	5.2, —	48.0	179.0
25/75	16	7.1	1270	1.7, 62.0	115.0	173.0
0/100	42	7.2	1400	63	115.0	173.0
P(3HB)/PCL[b]						
100/0	27	3	1500	—	—	—
90/10	26	11	1500	—	—	—
70/30	24	41	1300	—	—	—
50/50	23	631	1100	—	—	—
30/70	25	—	1300	—	—	—
0/100	16	920	500	—	—	—
P(3HB)/cellulose acetate butyrate[c]						
100/0	26	2	1679	—	—	—
90/10	27	2	2288	—	—	—
80/20	29	3	1672	—	—	—
70/30	20	2	1193	—	—	—
60/40	17	2	1154	—	—	—
50/50	13	7	592	—	—	—
0/100	6	1	1019	—	—	—

				Tg	Tc[g]	Tm[h]
P(3HB-co-3HV)/PLA[d]						
100/0	34	2	2593	-1.2, —	—	—, —
75/25	37	2	2453	18.3, 58.6	—	—
50/50	45	5	2357	16.7, 61.3	—	—
25/75	45	7	2212	16.3, 64.5	—	—
0/100	56	6	2175	—, 67.1	—	—
P(3HB-co-4HB)/PLA[e]						
100/0	14	2233	78	-10.7, —	—	—, —
30/70	29	214	1213	-14.0, 60.4	108.2	162.2, 168.1
20/80	38	317	1590	-13.9, 60.6	109.1	162.2, 168.2
10/90	43	273	1737	-13.1, 61.0	110.2	162.2, 168.5
5/95	49	96	1893	-6.9, 60.9	112	162.5, 168.5
0/100	65	6	1980	—, 61.5	114.1	163.1, 168.3

[a]Ref. 21
[b]Ref. 65
[c]Ref. 38
[d]Ref. 24
[e]Ref. 74
[f]Glass transition temperature
[g]Crystallisation temperature
[h]Melting temperature.

Shibata and co-workers[25] prepared different P(3HB-*co*-4HB)/PLA blends by varying the weight ratio of both polymers. The characteristics of the resulting blends with different P(3HB-*co*-4HB) copolymer content were then evaluated. A study of dynamic mechanical analysis (DMA) showed that increasing the P(3HB-*co*-4HB) copolymer content in the blends resulted in a reduction in T_g for the P(3HB-*co*-4HB) copolymer, with an unchanging T_g for the PLA. However, the T_gs for both polymers were not shifted toward each other. This suggested the immiscibility of the blends. P(3HB-*co*-4HB) copolymer was found evenly distributed in the PLA matrix with a clear and fine phase-separated-morphology under scanning electron micrograph (SEM) observation of the cryo-fractured surface of the blends after enzymatic removal of P(3HB-*co*-4HB) copolymer. With an increase in the P(3HB-*co*-4HB) copolymer content, the average particle size, indicated by the P(3HB-*co*-4HB) copolymer pores, was also increased, ranging from sub-micronic to micronic.

In both DMA and DSC analysis, the P(3HB-*co*-4HB) copolymer component in the blends showed lower T_g values as compared to the neat P(3HB-*co*-4HB) copolymer. With the increasing content of P(3HB-*co*-4HB) copolymer in the blends, the T_g was found to be further decreased. This was due to the immiscibility, resulting in the existence of a phase interface between PLA and P(3HB-*co*-4HB) copolymer. The chain end's enriched surroundings therefore enhanced the chain's mobility near the interface. Few studies have reported the depression of the T_g in an immiscible system and the observation was similar to the study of P(3HB-*co*-4HB)/PLA blends. As a result of the enhanced chain mobility near the interface, the incorporation of P(3HB-*co*-4HB) copolymer also enhanced the cold crystallisation of PLA by enhancing the nucleation of PLA. This was shown by a decrease in the cold crystallisation temperature. In terms of mechanical properties, addition of P(3HB-*co*-4HB) copolymer showed an increment of elongation at break, coupled with a decrease of modulus and tensile strength. Neat PLA was brittle and stiff and this was shown with an elongation at break of 5.5%. When 5 wt% of P(3HB-*co*-4HB) was blended into a PLA matrix, the elongation at break of the blend was improved, with a value of 96%. A further increment of P(3HB-*co*-4HB) content up to 20 wt% increased the elongation at break to a maximum value of 317%. After this point, the elongation at break started to reduce. Incorporation of P(3HB-*co*-4HB) copolymer, which is soft and elastomeric, was able to solve the brittleness of the neat PLA. Furthermore, it was expected that this was coupled with the reduction of tensile strength and modulus of the blends.[25]

The biodegradation of the P(3HB-*co*-4HB)/PLA blends was also evaluated by observing the surface morphology after enzymatic degradation. There was a slight increase of the enzymatic degradation of the blend with 10 wt% of P(3HB-*co*-4HB) copolymer as compared with neat PLA. A subsequent increment of P(3HB-*co*-4HB) copolymer content in the blend significantly enhanced the rate of enzymatic degradation. Addition of P(3HB-*co*-4HB) copolymer in the blends also resulted in an increase of the water absorption capability of the material.[25]

4.3.2 PHAs/PCL Blends

Chee and co-workers[28] performed viscometric studies on polymer-blend solutions of P(3HB) with PCL because it is a powerful method used to assess the miscibility of the components in an amorphous state. Viscometric analysis demonstrated that P(3HB) was immiscible with PCL. The phase behaviours of polymer blends were determined using Huggins' equation. Negative deviations from perfect behaviour were observed for the P(3HB)/PCL blend, suggesting repulsions between coils that lead to immiscibility. These observations were also in agreement with subsequent studies performed by Antunes and Felisberti[26] who reported that blends of P(3HB)/PCL with compositions of PCL that varied from 5 wt% to 30 wt% were immiscible as indicated by the formation of a system consisting of four distinct phases. DSC curves illustrated no changes in the shape or shifts in the peak related to T_m of P(3HB), suggesting that P(3HB) and PCL crystallised in two different phases. However, considerable degradation of P(3HB) and transesterification were not observed in these polymer blends obtained from the melting mixture; this might be because the polymers were processed for a short time and at a low temperature and in the absence of a catalyst.

The plasticizing effect of PCL had reduced the stiffness of injection-molded P(3HB)/PCL: 70/30 wt% in comparison to pure P(3HB). A small decrease in the tensile strength and modulus was observed. However, elongation at break increased tremendously from 18% to 87%. The DSC analysis of the P(3HB)/PCL: 70/30 wt% blend showed two T_gs, of −10.6 °C and −62.9 °C, which represented the P(3HB) matrix and PCL domains, respectively. The significant decrease in the T_g of P(3HB) (−0.8 °C) showed a partial miscibility of PCL in P(3HB). Injection-molding processing had resulted in the onset temperature and the maximum degradation temperature of P(3HB) decreasing from 283.4 °C and 307.2 °C to 280.2 °C and 301 °C, respectively. Addition of more thermally stable PCL had also slightly reduced the onset temperature of the P(3HB) degradation to 277.2 °C.[33]

Hinüber *et al.*[65] developed dimensionally stable hollow biocompatible and biodegradable fibers using a P(3HB/PCL): 70/30 wt% blend system through a melt-spinning method. This blend from a granulated mixture showed a significant difference in its bending stiffness compared with pure P(3HB) homopolymer (Figure 4.3). Elongation at break of this blend also exhibited an increment from 3% to 41% (Table 4.1). However, the blend showed elongation with necking but brittle-fractures were not observed. The necking behaviour might be attributed to the inhomogeneities in blend compositions and morphology. Elongation at break is not required to be high for the development of hollow fibers but it is necessary to prevent brittle fractures and obtain recognizable pliability. In that case, P(3HB/PCL): 70/30 wt% blend exhibited properties desired for regenerative medical therapies, for example as nerve guidance conduits.

It was reported by Jenkins and co-workers[32] that the T_g of the P(3HB-*co*-3HV)/PCL blend had two peaks indicating immiscibility of the blend.

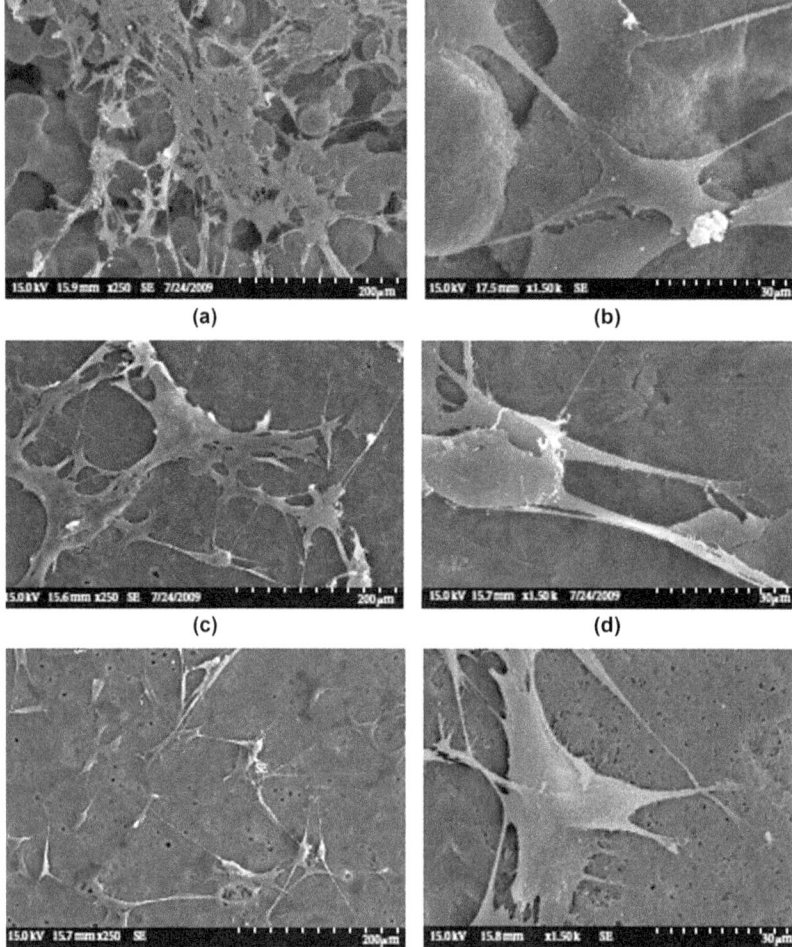

Figure 4.3   Image of extruded 'worm-like' structures to visualise flexibility: (left) P(3HB)/PCL 70/30 wt% (flakes) exhibits bendable properties, (right) pure P(3HB) shows a brittle fracture.[64]

Any miscibility that might have developed was lost on cooling. The immiscibility of the polymers was reported most probably because of the crystallisation of the P(3HB-*co*-3HV) copolymer followed by PCL, which led to phase separation between the blend. Qiu and co-workers[31] reported that the T_m of the blend remained almost the same as the individual polymers. This result suggested that the PCL and P(3HB-*co*-3HV) copolymer were immiscible. The presence of high P(3HB-*co*-3HV) content in the blend had a negative influence on the crystallisation of the PCL in the blend. Chun and Kim[29] reported a decrease of about 8 °C to 12 °C of the T_c of P(3HB-*co*-3HV)/ PCL blend in comparison to the T_c of pure P(3HB-*co*-3HV) copolymer. This decrease showed that PCL affected the crystallisation behaviour of

P(3HB-*co*-3HV) copolymer in P(3HB-*co*-3HV)/PCL blends. Qiu and co-workers[31] also reported that the addition of PCL to the blend reduced the crystallisation of P(3HB-*co*-3HV) copolymer. Therefore, the crystallisation of P(3HB-*co*-3HV) copolymer reduced after blending with PCL. That was attributed to the presence of PCL that suppressed the nucleation of P(3HB-*co*-3HV) in the blends.

4.3.3 PHAs/Cellulose Derivatives

Zhang *et al.*[35] reported an increase in the T_g of the annealed P(3HB)/ethyl cellulose (EtC) blend (100/0, 80/20, 60/40, 40/60, 20/80, and 0/100, wt%) from 44.6 °C to 70.4 °C with the increase of EtC content. However, a melt-quenched sample of EtC did not exhibit T_g. It was assumed that liquid crystals might exist below T_g because EtC is a thermotropic liquid crystal polymer. FT-IR analysis was also conducted to find the possible interaction between EtC and P(3HB). Hydrogen bonding of hydroxyls on the backbone of EtC makes it a rigid polymer with some degree of flexibility. P(3HB) possesses carbonyl groups that are proton acceptors along the main chain. A decrease in the absorption bands of hydroxyl groups in EtC was observed with an increase of P(3HB) content in the blends. However, the absorption bands of carbonyl groups at 1723–1724 cm^{-1} in P(3HB) were independent of the blend composition. It was proven that the hydrogen bonding of the hydroxyl groups of EtC was stronger than that of the hydroxyl group in EtC with the carbonyl group in P(3HB). Exothermic peaks corresponding to the crystallisation of P(3HB) were not detected in DSC traces of P(3HB)/EtC blends, thus it was concluded that a delay in the crystallisation of P(3HB) in the blends might be due to the EtC component. Polarizing optical micrographs (POM) of P(3HB) showed the growth of well-defined spherulites at 100 °C. However, no spherulite of P(3HB) was observed for the blends at a similar temperature. Depression of the spherulite growth was attributed to the dilution effect of the EtC component as well as the influence of melt miscibility on the primary nucleation processes. The results obtained had led to the conclusion that P(3HB)/EtC blends were miscible in an amorphous state. A SEM of the blends also showed the absence of phase separation as no distinct phase boundaries were observed.

Chan *et al.*[66] reported that the degradation rate increased with an increase of EtC content in the P(3HB)/EtC blends. Approximately 3.5% of the P(3HB) films' initial weight was lost after 105 days' incubation, whereas P(3HB)/EtC: 20/80 wt% lost about 12% of its initial weight. The influence of EtC on the biocompatibility of the P(3HB) films was compared using adult olfactory ensheathing cells (OECs). SEM illustrated that OECs readily attached to P(3HB) and P(3HB)/EtC: 80/20 wt% and exhibited healthy morphology with many filament extensions comparable to cells in an asynchronous control with the absence of biomaterials (Figure 4.4). Blending with EtC also exhibited changes in the surface structure of the P(3HB) polymer. The average surface roughness, R_a, of the P(3HB) films was increased with EtC content.

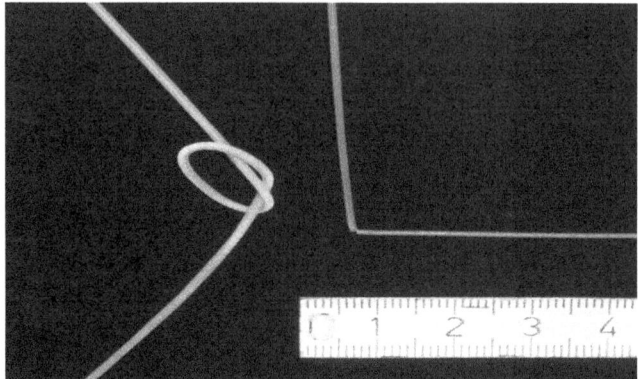

Figure 4.4　SEMs illustrating the attachment of olfactory ensheathing cells (OECs) on P(3HB)/EtC blend films with variations in EtC loadings after 24 hours' cultivation (a, d) P(3HB) films with 0%, (b, e) 20% and (c, f) 40% (w/w) ETC loadings; (a–c)×250, bar = 200 μm, (d–f)×1500, bar = 30 μm.[65]

The P(3HB)/EtC: 20/80 wt% blend had the most irregular surface with a significantly greater R_a value. Despite significant increases in R_a, no significant changes in morphology and proliferation of olfactory ensheathing cells were observed with the P(3HB)/EtC blends.

Blending of P(3HB) with cellulose had also been performed by Zhang and Deng[36] but with a different cellulose derivative, hydroxyethyl cellulose acetate (HECA). It was found that P(3HB) contents above 20% exhibited zero influence on the T_m of P(3HB) in the P(3HB)/HECA blend. However, an increase in the HECA component had resulted in the decrease of the melting enthalpy of the blends. The T_g of P(3HB) in the blend was constant at about 8 °C. FT-IR analysis showed that the two components did not exhibit specific interactions. It was found that the hydrogen bonding in HECA was so strong that the interaction between P(3HB) and HECA became very weak. The crystallisation of P(3HB) in the P(3HB)/HECA: 20/80 wt% blend was significantly affected by the HECA component. The phase transformation of HECA from an isotropic phase to a mesophase was easier than the crystallisation process, could be found at both lower and higher cooling rates and was almost independent of the P(3HB) component. However, crystallisation of P(3HB) was strongly dependent on the cooling rate and blend composition. Two separate transitions were found for P(3HB)/HECA (80/20), (60/40) and (40/60) blends during the DSC cooling run at a lower cooling rate, which corresponded to the crystallisation of P(3HB) and the phase transition of HECA from an isotropic phase to a mesophase in the blends, respectively.

Phase behaviour and crystallisation kinetics for the binary blend P(3HB)/ cellulose propionate (CP) were studied by Maekawa and co-workers.[37] A strong dependence of the measured T_g on composition was detected at high levels of CP. The Flory–Fox equation is one of the best equations to describe the dependence of T_g on composition in miscible blend systems. A good agreement between the measured and calculated values was detected using the Flory–Fox equation, which indicated that the system was miscible over the studied compositions (0 to 100% CP). A gradual increase in the T_g was observed for CP contents of 0% to 75%. However, the variation of T_g levelled off quite noticeably as the amount of P(3HB) in the system was decreased. A depression in the T_m was also observed with an increase in CP content, indicating miscibility between two components in the amorphous zones. The crystallisation rate of P(3HB) was significantly reduced with the addition of CP. Interestingly, the effect was more pronounced on the low-temperature side of the curves. This could be attributed to the restriction of segmental mobility of P(3HB), which disentangles from the CP chains in the homogeneous state at a lower temperature. P(3HB)/CP blend: 50/50 wt% exhibited tensile behaviour, which was intermediate between that of the two component polymers. It demonstrated a significant increase in ductility as compared with the homopolymers. An elongation at break of greater than 90% was achieved, which was due to a decrease in the crystallinity exhibited by the blend. This characteristic is usually observed in semi-crystalline-amorphous polymer blends.

Blending of P(3HB) with CP was also studied by Park and co-workers.[39] Unique orientation and crystallisation behaviours of P(3HB) in miscible blends with CP were investigated under uniaxial drawing and interpreted based on intramolecular nucleation and confined crystal growth. Formation of intramolecular nuclei with chain orientation perpendicular to the drawing directions was due to the fast relaxation of the P(3HB) component at the beginning of the annealing process. The crystal growth from these intramolecular nuclei proceeded along the long axis of narrow gaps formed by the extended ribbon-shaped CP chains. As a consequence, crystal lamellae with chain orientation perpendicular to the stretching direction were produced. Blend composition as well as annealing temperature were among the factors involved in determining the crystal growth mode of P(3HB) in P(3HB)/CP blends. Narrower gaps and slower crystallisation were observed for blends containing higher CP content. This was the origin of a higher degree of orthogonal crystalline orientation. Restraint of chain slippage and necking caused by the stiffness and high friction coefficient of the CP component had led to a decrease in the feasible maximum draw ratio with increasing CP content.

Wang and co-workers[38] studied miscibility, crystallisation behaviour, tensile properties and environmental biodegradability of P(3HB)/cellulose acetate butyrate (CAB) blends. The results of DSC, SEM, POM and wide-angle X-ray diffraction (WAXD) indicated that P(3HB) and CAB were miscible in the melt state. An increase in the CAB content decreased the T_m of the blend. The narrow processability window of the P(3HB) was broadened as the degree of its crystallinity was decreased. The toughness and ductility of the P(3HB) was also improved as indicated by an increase in elongation at break from 2.2% to 7.3%. The effect of pH on the degradation rates of P(3HB)/CAB blends was studied using buffer solutions of pH 1.0 and pH 13.0, respectively. A much slower degradation rate was observed in acidic media compared to alkaline media, suggesting that different routes were employed for the degradation depending on the pH of the surrounding environment. It was reported that neutral and acidic solutions produced more diffuse surface degradation, whereas degradation in alkaline solutions appeared to be more aggressive. However, degradation rates of P(3HB)/CAB blends were increased in both acidic and alkaline media in comparison to the pure P(3HB). An increase in the CAB content accelerated the degradation of blends in an acidic medium, whereas in an alkaline medium, the degradation rate decreased with an increase of CAB content.

P(3HB)/cellulose blend electroactive (EAP) polymer was studied by Zhijiang and co-workers.[34] It was reported that cellulose and P(3HB) had good miscibility and a uniform structure was observed for the P(3HB)/cellulose blend EAP. The existence of cellulose and P(3HB) in the blend was confirmed by FT-IR, X-ray diffraction (XRD) and DSC test results. Bending displacement, electrical power consumption and lifetime were evaluated under AC voltage under ambient conditions by changing the actuation voltage, frequency and time. An increase in the actuating voltage had increased the bending displacement of the P(3HB)/cellulose blend EAP. In comparison to pure P(3HB), a higher bending displacement of 1.32 mm was achieved at an actuating voltage of 7 V and a resonance frequency of 5 Hz. Although the electrical power consumption value was higher than that of pure cellulose EAP, it was still lower than the safety limit of microwaves, which might damage living organs. This novel P(3HB)/cellulose blend EAP had excellent durability and actuators because its bending lifetime for 10 hours only lost about 26% performance degradation at a room humidity level.

Hameed *et al.*[40] reported that as the concentration of cellulose increased in the blends, the T_g of the P(3HB-*co*-3HV) copolymer shifted slightly. At a higher concentration of cellulose, the T_g curve disappeared. This could be due to the partial miscibility of P(3HB-*co*-3HV) with cellulose. It was also observed that the T_m of the blend decreased with an increase in cellulose content in the blends. This was due to the reduction in the chemical potential. As for the mechanical properties, the Young's modulus was found to increase as the cellulose content in the blend increased. Elongation at break and tensile strength of the blend mainly depend on the composition of the blend. The tensile strength of the blend was found to be in between the tensile strength of P(3HB-*co*-3HV) copolymer and cellulose. As the cellulose content of the blend was increased, the tensile strength increased. This could be due to the higher strength of cellulose compared to P(3HB-*co*-3HV) copolymer. However, the elongation at break of the blend improved considerably compared to the individual components. As for the T_d, the onset T_d of the blend was in between the individual components. As the cellulose content in the blend increased, the T_d increased. This was due to the strong interaction between cellulose and P(3HB-*co*-3HV) copolymer, which was facilitated by the presence of ionic solutions.

4.3.4 PHAs/Starch Derivatives

According to Zhang and co-workers,[46] the T_g of P(3HB) in the blends was not affected by the addition of starch acetate (SA) as indicated by the constant temperature at about 9 °C, which was almost similar to pure P(3HB) homopolymer. Specific interactions between SA and P(3HB) were studied using FT-IR analysis. Absorption of hydroxyl groups in both P(3HB) and SA was indicative of intermolecular hydrogen bonding. In P(3HB)/SA blends, the O–H stretching band of SA presented almost in the same region at 3470 cm^{-1} as for pure SA while the absorption of carbonyl groups of P(3HB) in the blends was found to be independent of SA content at 1724 cm^{-1}. The C=O stretching band of the carbonyl groups of SA was also constant at 1748 cm^{-1}. This finding showed the absence of specific interactions occurring between P(3HB) and SA. SEM analysis illustrated the dispersion of the SA component in the P(3HB)/SA blends, which was far from uniform, indicating phase separation between P(3HB) and SA. Based on DSC, FT-IR and SEM, it was concluded that P(3HB)/SA blends were immiscible. A larger size of the SA phase was observed for the P(3HB)/SA: 40/60 wt% blend than for the P(3HB)/SA: 80/20 wt% blend. The crystallisation of P(3HB) from the melt was also observed using POM under isothermal conditions. Well-defined spherulites were observed at 80 °C for pure P(3HB). Crystallisation of P(3HB) was also observed in P(3HB)/SA: 80/20 wt% and 60/40 wt% blends as indicated by the spherulitic morphology. Although P(3HB) crystallised in the P(3HB)/SA: 40/60 wt% blend, spherulites were not found visually. Crystallisation of P(3HB) was not observed in the P(3HB)/SA: 20/80 wt% blend. The SA component also affected cold crystallisation. Shifting of the exothermal peaks to higher temperatures indicated the hindrance of mobile chains of P(3HB) above the T_g by the SA component.

Ismail *et al.*[47] studied the effect of starch content in the P(3HB) film matrix on its degree of swelling in water. Swelling in water and degradability are the most important characteristics for biodegradable materials. Polymer films were degraded by surface absorption of moisture and microorganisms. Both P(3HB) and P(3HB)/starch exhibited a gradual increase in water uptake during the first two days until it reached a plateau. The water uptake of the P(3HB) film was lower than that of the P(3HB)/starch blend, indicating that addition of starch decreased the hydrophobicity of P(3HB). The affinity for water is usually increased by the presence of hydroxyl groups on the surface of the starch granules. Therefore, the higher the starch content, the higher the water uptake in the P(3HB)/starch blends. The transport behaviour of water through the P(3HB)/starch blend is an important characteristic for biodegradability and applicability. The water resistance of the P(3HB)/starch blend was reduced with an increase in starch concentration, particularly above 10 wt%. This observation could be attributed to the capillary moisture formed by water on a starch basis.

The use of compatibilizers and plasticizers had resulted in the formation of a finer morphology of blend films, which might produce better mechanical properties. Good mechanical properties of P(3HB)/starch blends were observed at low concentrations of starch up to 10 wt%. However, a further increase of starch content decreased the stiffness of the blends due to weak adhesion between the P(3HB) matrix and starch particles. P(3HB)/starch: 90/10 wt% possessed a good combination of stiffness, strength, and toughness. The SEM micrograph of this blend showed relatively good dispersion and interfacial adhesion between starch granules and the P(3HB) matrix. The starch granules were also shown to be embedded in the P(3HB) matrix even after the rupture process.[47]

The biodegradation of the P(3HB)/starch blend was determined based on the weight loss measurement through utilization of P(3HB) and its blend by actinomycetes. The results indicated that the higher the starch content, the faster the rate of biodegradation. Degradation of the P(3HB)/starch blend was also affected significantly when the blends containing medium were supplemented with different carbon and nitrogen sources. A lower degradation of blend was observed for the medium supplemented with peptone, ammonium chloride and ammonium oxalate while an increase in the degradation was observed when calcium nitrate, potassium nitrate or ammonium molybdate were supplemented in the medium. The optimum conditions for the biodegradation of the P(3HB)/starch blend was at a pH of 7, temperature of 30 °C and medium supplemented with glucose or calcium acetate as carbon and nitrogen sources, respectively.[47]

The properties of P(3HB) when blended with two types of maize starch, Starch 1 (containing 70% amylose) and Starch 2 (containing 72% amylopectin) was examined by Zhang *et al.*[45] The SEM micrographs showed that starch acted as a filler for the P(3HB)/starch blend. The occurrence of intermolecular hydrogen bonding between P(3HB) and both types of maize starch was detected using FT-IR. In addition, the WAXD results revealed that the addition of starch affected the crystal structure of P(3HB). Starch 1 has stronger intermolecular hydrogen bonding with P(3HB) compared to Starch 2 due to its high-amylose content and linear structure. The existence of hydrogen bonding between the hydroxyl groups of starch and carbonyl groups of P(3HB) might inhibit chain scission degradation in P(3HB), which led to the improvement of the thermal stability, melt viscosity, and mechanical properties of the P(3HB)/starch blends. Higher melt shear viscosity, better mechanical properties and stability were observed for the blend containing Starch 1 compared to Starch 2 due to the stronger hydrogen bonding effect between P(3HB) and Starch 1.

Ramsay *et al.*[48] reported that addition of 50 wt% starch to P(3HB-*co*-3HV) copolymer resulted in a material that exhibited useful thermoplastic properties. The elongation at break increased as the amount of starch added decreased. The same trend was observed for Young's modulus. It was also reported that starch with altered surface properties would adhere more strongly to P(3HB-*co*-3HV) copolymer, thus increasing the strength of the blend.

Addition of glycerol as a plasticizer to the blend of P(3HB-*co*-3HV) polymer and starch produced the same trend where the elongation at break decreased as the amount of starch added increased reflecting the rigid nature of this polymer.[49] In another study by Reis *et al*,[52] the addition of starch content decreased the Young's modulus of P(3HB-*co*-3HV) polymer resulting in more flexible materials. The same trend was reported by Godbole *et al*.[67] and Thire *et al*.[68] Contradicting findings were reported by Koller and Owen[69] whereby the addition of starch increased the Young's modulus.

As the content of starch increased, the tensile strength decreased.[52] The same trend was observed by Koller and Owen.[69] Reis *et al*.[52] reported that the decrease in mechanical properties of the blend could be due to the low interfacial interaction between the components of the blend, which could lead to mechanical rupture. In another study by Thire *et al*,[68] it was a general trend that a better dispersed starch on the matrix would give better mechanical properties. This explained that the mechanical properties of the P(3HB-*co*-3HV) copolymer and starch blend were mainly contributed by the compatibility between hydrophobic P(3HB-*co*-3HV) copolymer and hydrophilic starch. As for the degradation, Ramsay *et al*.[48] showed an increased rate of degradation as the starch content increased. Loss of starch might be due to its solubility in water or enzymatic hydrolysis. Not only was it lost more rapidly than P(3HB-*co*-3HV) copolymer, but also its removal resulted in a larger exposed surface for the degradation of the P(3HB-*co*-3HV) copolymer. Since PHA is biodegradable, a decrease in starch content limited the rate rather than the extent of degradation.

The T_m of the blends depends on many factors. The T_m of P(3HB-*co*-3HV) copolymer in the blend was lower than that of the pure copolymer.[52] Matzinos *et al*.[70] reported that a small decrease in the T_m indicated that a phase separation had occurred. The melting enthalpy of the blend decreased with the increase of starch. Godbole *et al*.[67] reported an increase in the T_g value with increasing starch content. The same trend was observed by Seves *et al*.[51] using dynamic mechanical thermal behaviour (DMTA). Reis *et al*.[52] also reported a decrease in crystallisation enthalpy and T_c as the starch content increased. This indicated that blends had a lower degree of crystallinity compared to the pure P(3HB-*co*-3HV) copolymer. Any solid polymer can have amorphous, crystalline or both properties. In order to determine this, XRD was used by Reis *et al*.[52] There were no crystallinity value from the broad hump for the P(3HB-*co*-3HV)/starch blends. It was also observed that the crystallinity decreased as the starch content in the blend increased.

Zhang *et al*.[53] studied the thermal and mechanical properties of P(3HB-*co*-7%4HB)/corn starch blends. Besides that, the water absorption capability of the resulting blend films was also examined. DSC results showed that an increment of the corn starch content in the blend films had led to a decrease of crystallinity and also the T_m in comparison to pure P(3HB-*co*-7%4HB). The range of T_m was enlarged with the presence of double peaks. In terms of mechanical strength, an increase in the corn starch content led to a gradual decrease of the tensile and bending strength of the P(3HB-*co*-7%4HB)/corn starch blends. The elongation at break was found to be at its best when P(3HB-*co*-7%4HB) was blended with 20 phr corn starch, with an increment from 13.38% to 33.15%. The notched impact strength was the highest, with a value of 4.95 k Jm^{-2}, when 10 phr of corn starch was blended with the P(3HB-*co*-7%4HB) copolymer. The finding therefore suggested that blending with corn starch helped in reducing the crystallinity, which in turn increased the toughness of the blend material. In addition, it also reduced the intermolecular forces in the P(3HB-*co*-7%4HB) component, therefore the flexibility increased and subsequently resulted in a higher elongation. However, it was noted that an excessive amount of corn starch in the blend caused a decrease in the overall mechanical strength. The optimum tensile strength was found with the addition of 15 phr corn starch into the blend film. The water absorption capacity of the blend films was increased with an increase in corn starch, since corn starch is hydrophilic in nature.

4.3.5   PHAs/Chitosan

The thermal properties and crystallisation behaviour of P(3HB)/chitosan blends were studied by Ikejima *et al.*[54] using DSC and FT-IR. In comparison to pure P(3HB), P(3HB)/chitosan blend: 50/50 wt% exhibited lower T_m. A melting endotherm was not detected for P(3HB)/chitosan blends with a P(3HB) content of 10% and 25%. Suppression of the T_m of P(3HB) in the blend was attributed to the decrease of lamellar thickness of P(3HB). Intermolecular interactions between P(3HB) and the highly rigid chitosan molecules surrounding the P(3HB) molecules would make P(3HB) molecules in the blends inflexible and induce insufficient crystallisation compared to pure P(3HB). Degrees of crystallinity of P(3HB) were found to decrease in the P(3HB)/chitosan blend as determined using DSC and FT-IR analysis. This observation was attributed to the formation of intermolecular hydrogen bonds with the carbonyl groups of P(3HB) and the difference in the molecular mobility between P(3HB) and chitosan. Solid-state CP/MAS^{13}C NMR spectra confirmed the trapping of P(3HB) in the P(3HB)/chitosan blends. This trapping of the amorphous P(3HB) in the rigid chitosan amorphous phase was one of the factors that reduced the crystallinity of P(3HB) in the blends. Formation of intermolecular hydrogen-bonds in P(3HB)/chitosan blends was evidenced in the NMR spectra, but could not be deduced from FT-IR spectra.

Hu *et al.*[55] studied the biodegradation of P(3HB) grafted with chitosan *via* ozone treatment. Acrylic acid was grafted to ozone-treated P(3HB) and further grafted with chitosan (CS) *via* esterification. The biodegradability test was performed by placing the grafted membranes into a medium containing 1×10^5 cells/ml of *Alcaligenes faecalis* (ATCC 25094), followed by incubation at 37 °C for seven days. For P(3HB) and acetic acid grafted P(3HB), the bacteria grew from 1×10^5 cells/ml to 1.43×10^8 cells/ml. Although the surface of the P(3HB)/chitosan blend was more hydrophilic after grafting with chitosan, the growth rate of bacteria was reduced owing to their antibacterial activities. This had led to the reduction in the biodegradation rate of P(3HB)/chitosan blend.

DSC was used to evaluate the thermal properties and the miscibility of the P(3HB-*co*-3HV)/chitosan blend. For the T_m, the blend exhibited a general trend with respect to P(3HB). This indicated that there was a strong inter-molecular interaction between the carbonyl group in P(3HB) and the hydroxyl or amide group in chitosan. There was a single glass transition peak observed indicating that the blend of P(3HB-*co*-3HV) copolymer and chitosan were miscible in all proportions.[56]

Rennukka and Amirul[71] reported fabrication of P(3HB-*co*-4HB)/chitosan blends by adding chitosan powder (5, 10, 15 and 20 wt%) to P(3HB-*co*-4HB) with various 4HB molar fractions (10, 18 and 28 mol%). Since P(3HB-*co*-4HB) copolymer is hydrophobic in nature, it did not show any hydrophilicity regardless of various compositions of 4HB monomer. However, there was a significant improvement of the water adsorption capability when the chitosan content in the blends was increased from 5 to 20 wt%. Water up-take was influenced by the presence of the free hydrophilic groups in the blend films. The presence of chitosan induced the intra or/and inter-molecular network between the components in the blend films. Therefore, the hydrophilicity of the blend films increased.

FT-IR was used to determine the miscibility and examine the type of interactions formed in P(3HB-*co*-4HB)/chitosan blends. Initially, the absorbance spectrum for P(3HB-*co*-10%4HB), P(3HB-*co*-18%4HB) and P(3HB-*co*-28%4HB) were at 1746 cm^{-1}, 1756 cm^{-1} and 1759 cm^{-1} respectively, indicating the carbonyl ester (C=O) functional group. Whereas for chitosan film, three peaks corresponding to O–H stretching, C=O stretching (amide I) and N–H bending (amide II) were found at 3565 cm^{-1}, 1678 cm^{-1} and 1599 cm^{-1} respectively. For the P(3HB-*co*-10%4HB) blended with 5 and 20 wt% chitosan, the wavenumber of the carbonyl group of the P(3HB-*co*-4HB) component was found to be the same. However, the amino group for the chitosan was shifted to lower wavenumbers and closer to the carbonyl ester peak. A different trend was observed for P(3HB-*co*-18%4HB)/chitosan blends, however. The copolymer was shifted whereas there was a slight increase in the peaks of the amino group. For P(3HB-*co*-28%4HB), the carbonyl ester group occurred at a lower wavenumber and the amino group of chitosan was found to be increased. The difference in the functional groups of these components in the blend films as compared to the pure copolymer and chitosan films indicated the presence of intermolecular interactions between the carbonyl ester group of P(3HB-*co*-4HB) and the amino group of chitosan.[71]

The thermal properties of the P(3HB-*co*-4HB)/chitosan blends were studied in terms of the T_g, T_m and T_d. T_g was not detected for the blend with P(3HB-*co*-10%4HB) and P(3HB-*co*-18%4HB). For the blend film with P(3HB-*co*-28%4HB), there was an increase in the T_g from 55.9 °C to 70.8 °C upon increasing the chitosan content. Generally, there was a pattern of depression of the blend films when the chitosan content increased.[71]

The finding was supported by Wu[72] who reported that the presence of chitosan inhibited the movement of the polymer segment that in turn reduced the crystallinity. It was observed that the T_d decreased with the addition of 5 wt% chitosan but a further increment of the weight ratio increased the T_d, regardless of the variation of the 4HB molar fractions. It seemed that the addition of chitosan from 10 wt% led to a stronger interaction in the blends.

A surface morphology study of P(3HB-*co*-4HB)/chitosan blend films indicated an uniform and porous surface of the blends. By increasing the chitosan content, surface roughness was increased. This was followed by an increase in pore size. The authors suggested that P(3HB-*co*-28%4HB) with 20 wt% chitosan was a suitable implant material since the pore size was in a range that was significant for optimal bone growth. An *in vitro* microbial assay (disc diffusion method) of the blend films was carried out against *Escherichia coli* and *Staphylococcus aureus* which are Gram-negative and Gram-positive bacteria, respectively. For *E. coli*, the antimicrobial activity was increased with an increase in chitosan content. This was evidenced by an increase in the diameter of the clear zone from 6 to 11 mm. For *S. aureus*, the clear zone was maintained at 5 mm regardless of the concentration of chitosan in the blend films. The finding was also confirmed by the quantitative viable cell count method. Stronger antimicrobial properties were found towards *E. coli* due to the presence of the negatively charged cell membrane which was counteracted by the positively charged amino group. Therefore, it killed Gram-negative bacteria more effectively.[73]

4.4 Application of Blends of PHAs

Blending biopolymers with PHAs has gained a place in the current world trend concerned with environmental sustainability. The unending dependence on the limited resource of petroleum-based plastics has had a serious impact and has caused permanent damage to life on earth. Extensive efforts have been carried out, either to reduce their usage or to find an alternative resource, in an attempt to solve the problem. PHA itself is regarded as a potential alternative due to its biodegradability and biocompatibility. However, the blending of polymer with PHA can even advance and extend the applicability of pure PHA.[75] Blending of PHA with other biodegradable polymers is particularly prevalent when it comes to the creation of novel materials suitable for specific applications. Depending on the specific properties required, the blends could be tailor-made to incorporate some extra features or to improve the material properties currently present in particular PHAs.[76]

Peelman *et al.*[77] has extensively reviewed the application of bioplastics in food packaging. Food packaging has emerged progressively and is becoming more significant in the food industry. PLA and cellulose is common and widely used in packaging for both fresh and dried foods, such as yoghurt, chips, drinks, tomatoes and meat; whereas starch has been used to replace packaging that previously used polystyrene (PS). However, these polymers shared some common limitations, such as brittleness, difficult heat sealing ability, high oxygen and vapor permeability, poor mechanical strength and thermal instability. For example, starch is sensitive due to its high degree of hydrophilicity.[78] This subsequently hindered the practicality of the usage. Due to these drawbacks, it was suggested that by blending with PHAs such as P(3HB) and P(3HB-*co*-3HV), the functionality of the blends was further increased as compared to the pure ones. Therefore, the application of the resulting blends was more feasible and versatile in food packaging applications. Abdelwahab *et al.*[21] had developed a plasticized P(3HB)/PLA: 75/25 wt% blend using polyester plasticizer (Lapol 108) to create a new type of eco-friendly blend material suitable for single-use applications, such as fast food packaging.

Besides food packaging, the blends can be used in general packaging such as household items that are a part of daily consumerism. Homopolymer P(3HB) was extensively used as a raw material for these items. However, it exhibited properties such as high crystallinity, instability in the molten state and brittleness. Therefore, by blending with suitable biodegradable polymers, the applications were widened. It is more promising in practical applications such as kitchen films, cups, coats and sanitary napkins.[55,78] The blending was also expanded to the manufacturing of mulch films in agriculture.

An increasing demand for technological gadgets has also created an opportunity for blending. Blending is an economical yet useful way to tune the properties of the desired material. Therefore, blends made up of biodegradable polymers are attracting attention as materials for developing various electrical and electronic products, such as display panels, sensors and solar cells. Since they are biodegradable with zero conductivity, they can even be used as barrier materials for electromagnetic shielding, for a large range of products, including, scientific, medical, military and security products.[78]

The blending of PHAs was also applied to medical and biopharmaceutical applications. A few attempts were also carried out to test the applicability of P(3HB) and P(3HB-*co*-4HB) as biomaterials.[79,80] P(3HB-*co*-4HB) was found to be among the most suitable PHA biomaterials. This is because of the presence of eukaryote lipases that will hydrolyze the monomer into a natural metabolite.[81] Therefore, this copolymer has gained special attention when it comes to applications in the medical and biopharmaceutical fields. The blending plays its role when it comes to the creation of tissue engineering scaffolds. Since PHAs are hydrophobic in nature, this reduces the applicability in creating scaffolds. However, the incorporation of a hydrophilic biodegradable polymer has resulted in novel blends with improved wettability and water adsorption capacity. With an improved wettability, this enhanced and optimized cell growth on the surface of the blend, since cells tend to attach to a hydrophilic surface rather than a hydrophobic surface.[82] Development of hollow fibers by melt spinning using a P(3HB)/PCL blend was also reported by Hinüber *et al.*[65] in order to achieve a bendable hollow structure that is applicable in regenerative medical therapies, for instance as a nerve guidance conduit.

The blend created for tissue engineering scaffolds can also be used as a material for drug delivery. Reddy *et al.*[78] reported the development of a bone filler and drug delivery vehicle using a bionanocomposite, which was filled with bone morphogenetic protein, which stimulates bone formation. Besides that, a hydroxyapatite/chitosan nanocomposite was used in the controlled release of vitamins from the matrix. The blending of PHAs could also be used in a similar way since they share the same features, such as the biocompatibility and not causing adverse effects, which produce desirable clinical outcomes. Chan *et al.*[66] developed P(3HB)/EtC blends with a significant reduction in crystallinity, which enhanced their degradation without compromising material properties and biocompatibility. It was suggested that with the added control of annealing, tailor-made P(3HB)-based biomaterials could be achieved by blending to support biomedical device fabrication such as nerve conduits for neuronal regeneration as well as the development of tissue engineering matrices.

4.5   Conclusions

Biodegradable polymers are making their mark in a plethora of fields due to their good biodegradability and biocompatibility. For the last two decades, these polymers have been the focus of many researchers because the advancement of these polymers could contribute to sustainable development as they offer an advantage from a cost effectiveness point of view, with a wide range of disposal options with a lower environmental impact. Blending has attracted the attention of researchers mainly because it is more economical and can produce polymers that are hybrids of two polymer components with interesting properties. At present, many PHA blends have been studied and produced with particular characteristics able to cover a large spectrum of specific needs. PHA blends will remain competitive if they can be developed at a lower cost with properties tailored for particular applications. This review outlines the significance of the research and development that has been undertaken in the field of polymer blends using P(3HB), P(3HB-*co*-3HV) and P(3HB-*co*-4HB) with PLA, PCL, cellulose, starch and chitosan. Enhancement of PHAs' properties, particularly as biopolymers, has been successfully obtained by modifying the PHAs' physiochemical and material properties through blending, usually *via* solvent-casting and melt-compounding methods. However, these properties are more favourable for environmental applications with limited success in biomedical scenarios. Intensive research needs to be conducted to improve compatibilizing mechanisms, surface modification and advanced processing techniques in an attempt to obtain ideal polymeric blends and composites from biodegradable polymers.

References

1. L. Avérous and E. Pollet, Biodegradable Polymers, in *Environmental Silicate Nano-Biocomposites (Series: Green Energy and Technology)*, Springer-Verlag, Berlin, Heidelberg, 2012, pp. 13–39.
2. L. Yu, K. Dean and L. Li, *Prog. Polym. Sci.*, 2006, **31**, 576.
3. E. V. Prut and A. N. Zelenetskii, *Russ. Chem. Rev.*, 2011, **70**, 65.
4. L. A. Utracki, *Polymer blends handbook*, Kluwer Academic Publishers, USA, 2002, pp. 1–1422.
5. E. Akaraonye, T. Keshavarz and I. Roy, *J. Chem. Technol. Biotechnol.*, 2010, **85**, 732.
6. S. Philip, T. Keshavarz and I. Roy, *J. Chem. Technol. Biotechnol.*, 2007, **82**, 233.
7. M. Koller, I. Gasser, F. Schmid and G. Berg, *Eng. Life Sci.*, 2011, **11**, 222.
8. C. S. K. Reddy, R. Ghai, Rashmi and V. C. Kalia, *Bioresour. Technol.*, 2003, **87**, 137.
9. P. Y. Tian, L. Shang, H. Ren, Y. Mi, D. D. Fan and M. Jiang, *Afr. J. Biotechnol.*, 2009, **8**, 709.
10. J. G. C. Gomez, B. S. Méndez, P. I. Nikel, M. J. Pettinari, M. A. Prieto and L. F. Silva, *Making green polymers even greener: Towards sustainable production of polyhydroxyalkanoates from agroindustrial by-products*, ed. M. Petre, *Adv. Appl. Biotechnol.*, InTech, Croatia, 2012, pp. 41–62.
11. A. Steinbüchel and T. Lütke-Eversloh, *Biochem. Eng. J.*, 2003, **16**, 81.
12. G. Q. Chen, *Chem. Soc. Rev.*, 2009, **38**, 2434.
13. A. J. Anderson and E. A. Dawes, *Microbiol. Rev.*, 1990, **54**, 450.
14. S. J. Park, T. W. Kim, M. K. Kim, S. Y. Lee and S. C. Lim, *Biotechnol. Adv.*, 2012, **30**, 1196.
15. C. T. Nomura, T. Tanaka, Z. Gan, K. Kuwabara, H. Abe, K. Takase, K. Taguchi and Y. Doi, *Biomacromolecules*, 2004, **5**, 1457.
16. R. Babu and T. Woods, *SPE Plast. Res. Online*, 2011, pp. 1–2.
17. M. Avella, E. Martuscelli and M. Raimo, *J. Mater. Sci.*, 2000, **35**, 523.
18. L. Zhang, C. Xiong and X. Deng, *Polymer*, 1996, **37**, 235.
19. I. Ohkoshi, H. Abe and Y. Doi, *Polymer*, 2000, **41**, 5985.
20. M. Zhang and N. L. Thomas, *Adv. Polym. Technol.*, 2011, **30**, 67.
21. M. A. Abdelwahab, A. Flynn, B. S. Chiou, S. Imam, W. Orts and E. Chiellini, *Polym. Degrad. Stab.*, 2012, **97**, 1822.
22. M. R. Nanda, M. Misra and A. K. Mohanty, *Macromol. Mater. Eng.*, 2011, **296**, 719.
23. H. Zhao, Z. Cui, X. Sun, L. H. Turng and X. Peng, *Ind. Eng. Chem. Res.*, 2013, **52**, 2569.
24. I. Zembouai, M. Kaci, S. Bruzaud, A. Benhamida, Y. M. Corre and Y. Grohens, *Polym. Test.*, 2013, **32**, 842.
25. M. Shibata, Y. Inoue and M. Miyoshi, *Polymer.*, 2006, **47**, 3557.
26. M. C. M. Antunes and M. I. Felisberti, *Polímeros*, 2005, **15**, 134.
27. D. Lovera, L. Márquez, V. Balsamo, A. Taddei, C. Castelli and A. J. Müller, *Macromol. Chem. Phys.*, 2007, **208**, 924.

28. M. J. K. Chee, J. Ismail, C. Kummerlöwe and H. W. Kammer, *Polymer*, 2002, **43**, 1235.
29. Y. S. Chun and W. N. Kim, *Polymer*, 2000, **41**, 2305.
30. K. Wessler, M. H. Nishida, J. da Silva, A. P. T. Pezzin and S. H. Pezzin, *Macromol. Symp.*, 2006, **245–246**, 161.
31. Z. Qiu, W. Yang, T. Ikehara and T. Nishi, *Polymer*, 2005, **46**, 11814.
32. M. J. Jenkins, Y. Cao, L. Howell and G. A. Leeke, *Polymer*, 2007, **48**, 6304.
33. M. A. T. Duarte, R. G. Hugen, E. S. Martins, A. P. T. Pezzin and S. H. Pezzin, *Mater. Res.*, 2006, **9**, 25.
34. C. Zhijiang, H. Chengwei and Y. Guang, *Carbohydr. Polym.*, 2012, **87**, 650.
35. L. Zhang, X. Deng and Z. Huang, *Polymer*, 1997, **38**, 5379.
36. L. Zhang and X. Deng, *Polymer*, 1997, **38**, 6001.
37. M. Maekawa, R. Pearce, R. H. Marchessault and R. S. J. Manley, *Polymer*, 1999, **40**, 1501.
38. T. Wang, G. Cheng, S. Ma, Z. Cai and L. Zhang, *J. Appl. Polym. Sci.*, 2003, **89**, 2116.
39. J. W. Park, Y. Doi and T. Iwata, *Macromolecules*, 2005, **38**, 2345.
40. N. Hameed, Q. Guo, F. H. Tay and S. G. Kazarian, *Carbohydr. Polym.*, 2011, **86**, 94.
41. D. F. Gilmore, R. C. Fuller, B. Schneider, R. W. Lenz, N. Lotti, N and M. Scandola, *J. Environ. Polym. Degrad.*, 1994, **2**, 49.
42. N. Lotti and M. Scandola, *Polym. Bull.*, 1992, **29**, 407.
43. J. Fang and P. Fowleer, *J Food, Agric. Environ.*, 2003, **3 and 4**, 82.
44. L. Yu, *Biodegradable polymer blends and composites from renewable resources*, John Wiley & Sons, Hoboken, New Jersey, 2009, pp. 1–191.
45. M. Zhang and N. L. Thomas, *J. Appl. Polym. Sci.*, 2010, **116**, 688.
46. L. Zhang, X. Deng, S. Zhao and Z. Huang, *Polym. Int.*, 1997c, **44**, 104.
47. A. M. Ismail and M. A. B Gamal, *J. Appl. Polym. Sci.*, 2010, **115**, 2813.
48. B. A. Ramsay, V. Langlade, P. J. Carreau and J. A. Ramsay, *Appl. Environ. Microbiol.*, 1993, **59**(4), 1242.
49. N. F. Magalhaes and C. T. Andrade, *Polímeros*, 2013, **23**, 366.
50. M. F. Koenig and S. J. Huang, *Polymer*, 1995, **36**, 1877.
51. A. Seves, P. L. Beltrame, E. Selli and L. Bergamasco, *Angew. Makromol. Chem.*, 1998, **260**, 65.
52. K. C. Reis, J. Pereira, A. C. Smith, C. W. P. Carvalho, N. Wellner and I. Yakimets, *J. Food Eng.*, 2008, **89**, 361.
53. J. Zhang, X.-P. Lu, X.-Y. Ren and X. Wen, *J. Tianjin Univ. Sci. Tech.*, 2010, **25**, 34.
54. T. Ikejima, K. Yagi and Y. Inoue, *Macromol. Chem. Phys.*, 1999, **200**, 413.
55. S.-G. Hu, C.-H. Jou and M.-C. Yang, *J. Appl. Polym. Sci.*, 2003, **88**, 2729.
56. M. K. Cheung, K. P. Y. Wan and P. H. Yu, *J. Appl. Polym. Sci.*, 2002, **86**, 1253.
57. Y. Saito, S. Nakamura, M. Hiramitsu and Y. Doi, *Polym. Int.*, 1996, **39**, 169.
58. M. N. V. R. Kumar, *React. Funct. Polym.*, 2000, **46**, 1.

59. N. Koyama and Y. Doi, *Polymer*, 1997, **38**, 1589.
60. C. Vogel, E. Wessel and H. W. Siesler, *Biomacromolecules*, 2008, **9**, 523.
61. I. Noda, M. M. Satkowski, A. E. Dowrey and C. Marcott, *Macromol. Biosci.*, 2004, **4**, 269.
62. I. M. Neill and H. Leiper, *Polym. Degrad. Stab.*, 1985, **11**, 309.
63. E. Richards, R. Rizvi, A. Chow and H. Naguib, *J. Polym. Environ.*, 2008, **16**, 258.
64. S. Modi, K. Koelling and Y. Vodovotz, *J. Appl. Polym. Sci.*, 2012, **124**, 3074.
65. C. Hinüber, L. Häussler, R. Vogel, H. Brünig, G. Heinrich and C. Werner, *Polym. Lett.*, 2011, **5**, 643.
66. R. T. H. Chan, C. J. Garvey, H. Marçal, R. A. Russell, P. J. Holden and J. R. Foster, *Int. J. Polym. Sci.*, 2011, **2011**, 1.
67. S. Godbole, S. Gote, M. Latkar and T. Chakrabarti, *Bioresour. Technol.*, 2003, **86**, 33.
68. R. M. S. M. Thire, T. A. A. Ribeiro and C. T. Andrade, *J. Appl. Polym. Sci.*, 2006, **100**, 4338.
69. I. Koller and A. J. Owen, *Polym. Int.*, 1996, **39**, 175.
70. P. Matzinos, V. Tsuki, A. Kontoyiannis and C. Panayiotou, *Polym. Degrad. Stab.*, 2002, 77, 17.
71. M. Rennukka and A. A. Amirul, *Polym. Bull.*, 2013, **70**, 1937.
72. C. S. Wu, *Polymer*, 2005, **46**, 147.
73. C. M. Murphy, M. C. Haugh and F. J. O'Brien, *Biomaterials*, 2009, **31**, 461.
74. L. Han, C. Han, H. Zhang, S. Chen and L. Dong, *Polym. Compos.*, 2012, **33**, 850.
75. R. M. Rasal, A. V. Janorkar and D. E. Hirt, *Prog. Polym. Sci.*, 2010, **35**, 338.
76. T. Gerard and T. Budtova, *Preparation and characterization of poly-hydroxyalkanoates (PHA) and polylactide (PLA) blends*, presented in part at the 27th World Congress of the Polymer Processing Society, Polymer Processing Society, Morocco, 2011.
77. N. Peelman, P. Ragaert, B. De Meulenaer, D. Adons, R. Peeters, L. Cardon, F. Van Impe and F. Devlieghere, *Trends Food Sci. Technol.*, 2013, **32**, 128.
78. M. M. Reddy, S. Vivekanandhan, M. Misra, S. K. Bhatia and A. K. Mohanty, *Prog. Polym. Sci.*, 2013, **38**, 1653.
79. R. C. Young, M. Wiberg and G. Terenghi, *Br. J. Plast. Surg.*, 2002, **55**, 235.
80. Y. W. Wang, F. Yang, Q. Wua, Y. Cheng, P. H. F. Yu and J. Chen, *Biomaterials*, 2005, **26**, 755.
81. D. P. Martin and S. F. Williams, *Biochem. Eng. J.*, 2003, **16**, 97.
82. U. Rao, R. Kumar, S. Balaji and P. K. Sengal, *J. Bioact. Compat. Polym.*, 2010, **25**, 419.

Nanocomposites of Polyhydroxyalkanoates (PHAs)

A. M. GUMEL AND M. S. M. ANNUAR*

Institute of Biological Sciences, Faculty of Science, University of Malaya, 50603 Kuala Lumpur, Malaysia
*Email: suffian_annuar@um.edu.my

5.1 Introduction

Polyhydroxyalkanoates (PHAs) are a group of polyhydroxyesters with diverse structures. The composition of these polyesters may include 3-, 4-, 5- and 6-hydroxyalkanoic acid monomers that are accumulated by different bacterial species (native or mutant) under nutrient(s)-deprived conditions but abundant carbon sources. A wide range of carbon sources could serve as substrate(s) for microbial PHA accumulation but in terms of its mass production, the use of renewable resources such as fatty acids is very favorable.[1–3] The accumulated PHAs serve as a carbon and energy reserve within the microorganisms.[4] A range of PHAs comprising both copolymers and block copolymers has been produced using different processes including fermentation and enzymatic catalysis, leading to an accumulated PHA content of as much as 90% of the microbial dry mass.[5] Moreover, alternative production schemes based on genetically modified plants and yeast are currently gaining momentum and may become the preferred route for PHA production.[5] Unlike petrochemical-derived plastics, PHAs are biodegradable, and are biocompatible with gas-barrier properties almost similar to those of polyvinyl chloride and polyethylene terephthalate.[6] These combinations of excellent physico–chemical properties warrant the increasing

RSC Green Chemistry No. 30
Polyhydroxyalkanoate (PHA) Based Blends, Composites and Nanocomposites
Edited by Ipsita Roy and Visakh P M
© The Royal Society of Chemistry 2015
Published by the Royal Society of Chemistry, www.rsc.org

commercial exploitation of biopolymers in different niche applications spanning from biomedical, packaging, automotive, infrastructure, and aerospace to military applications.[5]

Unfortunately, despite their promising commercial potential, most of the produced PHAs, especially those with a higher monomeric composition of 3-hydroxybutyric acid, were reported to exhibit brittleness, low heat distortion temperature, and poor gas-barrier properties, limiting processing malleability and ductility.[7,8] This in turn justified the current concern regarding the durability of the materials based on their strength, shelf lifetime, replacement and maintenance cost. Fortunately, nanoreinforcement of these kinds of biopolymers into nanocomposites to improve their quality and properties has already been proven thereby extending their applications in aggressive environments were the neat polymers may fail.[9] It is possible to overcome these shortcomings by modifying the biopolymers through nanoreinforcements using nanofillers such as layered silicates or phyllosilicates.[9] Therefore, currently, nanoreinforcements and the resulting nanocomposite(s) are a focus of intense research to improve the qualitative performance of PHAs.

The application of nanoparticles (nanofillers) as composite agents is attractive because the nanofillers not only improve polymer crystallization but also the gas-barrier, thermo-mechanical and physico–chemical properties, surpassing those of the native biopolymers or its conventional microcomposites.[10] Such nanocomposites are expected to be significant due to the ease of nanoscale dispersion (≈ 1 nm) with a minute amount of nanofiller ($\approx 5\%$ w/w), resulting in a higher aspect ratio and extendable surface area.[7] It is obvious that the incorporation of nanofillers in PHA nanocomposites expands their applications by improving the performance of the polymer significantly. The most commonly used nanofillers include silylated kaolinite,[11] multiwall carbon nanotubes,[12] bioactive glass,[13] organophilic montmorillonite,[14] nanoclay,[15] cellulose nanocrystals,[16] layered double hydroxides,[17] cobalt–aluminium layered double hydroxides,[18] *etc.*

Of all the nanofillers studied, clays such as layered silicates of montmorillonite, saponite and hectorite were found to be the most suitable due to the presence of an expanded interlayer van der Waals gap that can intercalate polymeric molecules.[9,19] However, the opposing polarity between the hydrophilic silicates and the hydrophobic polymer matrix mostly results in a lack of affinity between the silicate and the polymer, making it difficult to achieve a homogeneous mixture. For example, mostly hydrated Na^+ or K^+ ions are said to be distributed on the neat layered silicate surface.[9] This kind of surface is only miscible with hydrophilic polymers such as polyvinyl alcohol, polyamide, polyacrylamide, cellulose, proteins, polyethylene glycol and ethoxylated graft polymers.[20] Unmodified silicate nanolayers result in weaker nanocomposites with higher agglomeration of the nanofillers due to a poor physical attraction between the organic and inorganic components. Therefore, to render the hydrophilic surface of the silicates nanolayers miscible with the hydrophobic polymer, a modification *via* surface chemistry manipulation has to be made.[19] This is generally achieved by ion-exchange reactions using cationic surfactants such as primary, secondary, tertiary, and quaternary alkylammonium or alkylphosphonium.[9] The presence of these cations in the organosilicates is suggested to have a reducing effect on the inorganic host's surface energy, thereby improving the wetting property of the polymer matrix, resulting in a large interlayer spacing.[9]

Consequently, layered silicate nanofillers are said to be indispensable in PHA nanocomposites technology.[19] In general, the physico–mechanical properties of nanocomposite materials are greatly influenced by the degree of mixing between the two phases, and the quality of structure intercalation depends on the extent to which the organic component and the inorganic components are able to interact and thus be made compatible. In the following sections, different methods of PHA nanocomposite production, characterization and applications are discussed.

5.2   Production of PHA Nanocomposites

A number of methods have frequently been employed in the production of nanocomposite materials. These include solution intercalation,[11] melt intercalation,[21] polymerization,[22] sol–gel,[23] deposition,[24] magnetron sputtering,[25] laser,[26,27] ultrasonication,[28] supercritical fluid,[29,30] *etc*. In PHA nanocomposite fabrication, solution intercalation and melt intercalation methods are the most widely explored procedures.[3,7] However, use of *in situ* intercalative polymerization, supercritical fluids and electrospinning are shown to be promising and emerging techniques. The performance and quality of a nanocomposite depends on how well the nanofillers disperse or blend into the matrix. Therefore, these methods constitute different strategies to improve the composites' thermo–mechanical and physico–chemical properties by enhancing efficient interactions between the nanofiller and the polymer matrices.

When using nanofillers such as layered silicate, different morphologies could be observed depending on how well the nanofillers were dispersed in the matrix. Poor dispersion of nanofillers often results in aggregate (tactoids) morphologies (Figure 5.1).

However, interactions between the nanofillers with some of the extended polymer chains result in an intercalated nanocomposite with an alternate arrangement of polymer and layered nanofiller (Figure 5.1). Absolute and homogeneous dispersion of nanofillers in the polymer matrix gives a nanocomposite with exfoliated morphological characteristics (Figure 5.1). Previously, this ordered exfoliation was thought to be induced by steric interactions.[10,31] However, some researchers, such as Zhang *et al.*[32] suggested that the dispersion of the polymeric side chain into the nanofillers' layered spaces was the main cause of exfoliation.

The surface of hydrophilic nanofillers is usually functionalized (modified) so as to promote a homogeneous dispersion throughout the polymer, and to enhance interactions within the polymeric matrix, resulting in improved

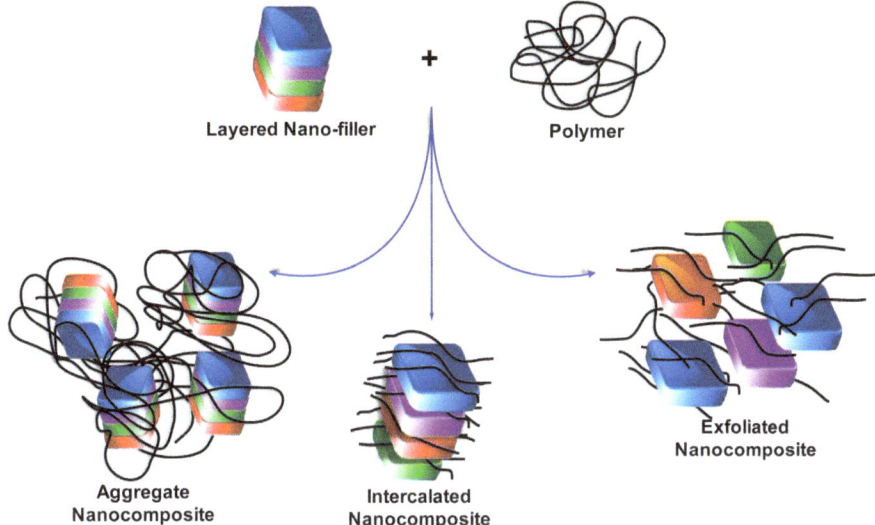

Figure 5.1 Schematic illustration of polymer/layered nanofiller composite morphologies.

thermo–mechanical and physico–chemical properties of PHAs.[11] It has been reported that the addition of a surfactant as a modifier in the nanocomposite preparation offers sufficient excess enthalpy for complete and homogeneous dispersion of the nanofiller into the polymer matrix.[10] The applications of several surfactants/modifiers have been investigated in such processes (Table 5.1). The type of polymeric matrix often dictates the type of surfactant and/or modifier to be used for better composition. For example, in a nanocomposite using a hydrophobic polymer matrix and layered hydrophilic silicate nanofillers, an alkyl ammonium surfactant has been found to be the best modifier.[10] This is because the alkyl ammonium cations could undergo an ion exchange reaction converting the hydrophilic silicate surface to a compatible hydrophobic surface, while at the same time providing a functional group that can interact with the polymer matrix.[33]

5.2.1 Solution Intercalation

The solution intercalation (solution casting) method is based on the solvation of the polymer matrix into the solvent (*e.g.* chloroform, toluene, dichloromethane, *etc.*), and the swelling of the nanofillers within the solvent.[35] In this method, the nanofillers are normally allowed to swell in the solvent by incubation for a specified period of time.[3] Then the polymer is added to solubilize and intercalate with the layered nanofillers upon homogeneous mixing at a specific temperature. Depending on the experimental design, the resulting mixture may be allowed to stand for some time, usually 4–5 days under regular periodic shaking before evaporating the

Table 5.1 Some commonly used surfactants/modifiers for nanofiller modification in PHA nanocomposites.[a]

Polymer	Nanofiller	Surfactant/modifier	Preparation method	Ref.
PHBHHx	Layered silicate (clay 25A)		Melt intercalation	34
PHBHV	Montmorillonite		Solution intercalation	35
PHBHHx	Kaolinite		Sol–gel/ solution casting	11
PHB	Bioglass		Salt leaching	36
PHB/PHBHV	Clay		Solution casting	37
PHB/PHBHV	Clay		Solution casting	37
PHB/PHBHV	Clay		Solution casting	37
PHB	Synthetic fluoromica		Melt extrusion	38
PHB	Montmorillonite		Melt extrusion	38
PHB	Cloisite® 30B		Melt-mixing	39

[a]HT = hydrogenated tallow, T = tallow, PHB = poly(3-hydroxybutyrate), PHBHHx = poly(3-hydroxybutyrate-*co*-3-hydroxyhexanoate), PHBHV = poly(3-hydroxybutyrate-*co*-3-hydroxyvalerate).

solvent.[35] However, in another procedure, the solvent is immediately evaporated after mixing.[32] Upon solvent evaporation, the intercalated polymeric materials give a nanocomposite that is normally dried *in vacuo*.

The thermodynamics involved in this process were explained previously.[40] In brief, before the solubilized polymer intercalates the layered nanofillers, the solvent molecules that occupy the spaces in between have to be displaced; hence a negative variation in the Gibbs free energy change is required. The driving force for this process is hypothesized to be due to entropy gained by desorption of solvent molecules, compensating the decreased entropy of the confined intercalated polymeric chain.[3,7]

In the solution casting method, the polarity of the solvent is critical in facilitating the intercalation of polymers into the layered nanofillers.[41] A strongly polar solvent causes excessive swelling of the silicate's hydrophilic surface, resulting in fracture or cracking of the materials and low polymer solvation. A highly hydrophobic solvent influences the solvation of the polymer but leads to poor silicate nanofiller swelling. This warrants the use of surfactants and/or modifiers in order to compromise the limitations that arise from these disparities.

Polyhydroxyalkanoates, such as poly-3-hydroxybutyrate,[42] poly-3-hydroxybutyrate-*co*-3-hydroxyvalerate,[15] and poly-3-hydroxybutyrate-*co*-3-hydroxyhexanoate,[11,21] were intercalated using this method. Ten *et al.*[43] produced evenly exfoliated nanocomposites of PHBV/cellulose nanowhiskers, with a well-defined distribution of the nanofillers within the polymer matrix (Figure 5.2).

Observing the effects of nanofiller loading during Nodax™ (PHBHHx)/layered nanosilicate fabrication in chloroform solution casting, Zhang *et al.*[32] found that the addition of nanofillers (≤5%) increases thermal properties such as Young's modulus and the tensile strength of the nanocomposites as compared to those of the pure polymer. On the other hand, a

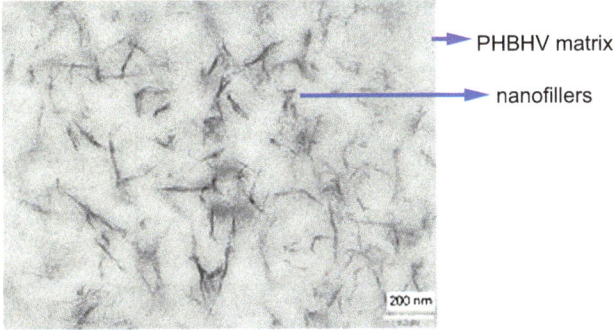

PHBHV matrix

nanofillers

200 nm

Figure 5.2 TEM micrograph showing exfoliated distribution of cellulose nanowhiskers in a PHBHV matrix nanocomposite prepared by the solution intercalation method.
Reprinted from Ten *et al.*[43] with permission from Elsevier.

higher increase in nanofiller loading (>5%) does not improve the mentioned mechanical properties.[32]

Zhijiang et al.[42] reported an improvement in the mechanical properties of a PHB nanocomposite made of bacterial cellulose nanofibrils that was pre-pared by the solution casting method. In addition, they found that the nanocomposite showed better biocompatibility and mechanical properties than pure PHB based on cell-adhesion analysis using Chinese hamster lung (CHL) fibroblast cells and stress strain tests, respectively.[42] In comparison to pure PHB, the nanocomposite of PHB/bacterial cellulose was observed to exhibit about a 202% increase in tensile stress and a 2.2-fold increase in elongation to break, respectively (Figure 5.3).[42]

The same solution casting method was employed in the authors' labora-tory to study the degradation behavior and thermo–mechanical properties of nanocomposites based on medium-chain-length PHA/carbon nanofibers (CNF) in an ultrasound assisted process. At CNF loading ≤10% w/w, a cor-relative increase in the nanofiller dispersion with increasing sonication power output and exposure time was observed. This led to the formation of an exfoliated nanocomposite. In contrast, when the CNF nanofiller loading was increased beyond 10% w/w, a nanocomposite with agglomer-ated morphology was observed. Ten et al.[44] studied the isothermal crystal-lization kinetics of poly(3-hydroxybutyrate-*co*-3-hydroxyvalerate) (PHBV)

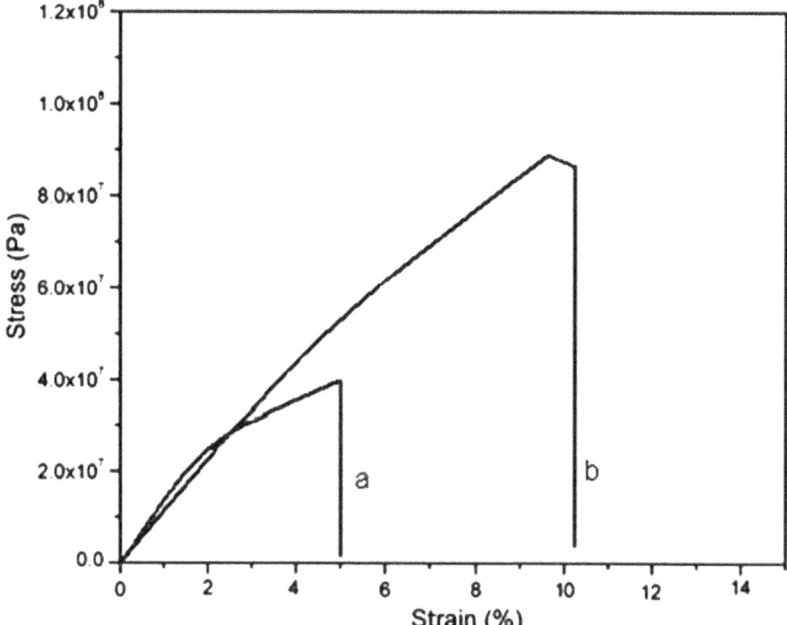

Figure 5.3   Comparison of tensile test results on (a) PHB and (b) a PHB/bacterial cellulose nanocomposite.
Reprinted from Zhijiang et al.[42] with permission from Elsevier.

nanocomposites containing cellulose nanowhiskers (CNWs). The researchers studied the effects of CNWs concentration and temperature on the crystallization rate and crystallinity of the PHBV nanocomposite using the Avrami equation (eqn (5.1)),

$$1 - X = e^{-Kt^n} \tag{5.1}$$

where X is the degree of crystallization, K is the nucleation growth rate, t is time and n is the model integer from 1 to 4.[45,46]

Their results suggested that the crystallinity of the PHBV nanocomposite is highly modulated by the dual effects of nucleation and confinement induced by the CNWs' concentration. They reported that at low CNW concentrations, a heterogeneous nucleation was favored, which in turn increased PHBV crystallization. On the other hand, the crystallization rate started to decrease when the confinement effect of CNWs outweighed their nucleation effect.[44] This effect could be explained as follows: a higher CNW concentration results in an increase in interfacial area and nucleation sites. This accelerates the crystallization rate of the PHBV. However, an excessive increase of CNW loading beyond the agglomeration point leads to the confinement of the PHBV chain. The confinement effect retards PHBV crystallite growth by hindering the polymer chain diffusion to the growing crystallite.[44]

Although the solution intercalation method is reported to be a favorable method for highly hydrophobic polymer intercalation into layered nanofillers,[33] at present its industrial scale-up is not feasible. This is mostly due to the large quantity of organic solvent needed, incurring higher production costs, and the environmental concerns due to solvent pollution.

5.2.2 Melt Intercalation

As an alternative to avoid the excessive use of organic solvents during the solution intercalation process, the melt intercalation method gained momentum in the early 1990s.[47] This method involves mixing the polymer powder with the nanofiller followed by pressing the mixture into pellets, followed by heating in a thermal chamber at a specific temperature under continuous mixing. Using an ammonium-modified montmorillonite clay (Cloisite 30B), Mook Choi *et al.*,[48] for the first time, described the production of an intercalated PHBHV-clay nanocomposite using melt extrusion in a Brabender mixer at 165 °C, with an agitation rate of 50 rpm for 15 min. The same melt extrusion was employed by Maiti *et al.*[38] to produce a well dispersed PHB/layered silicate nanocomposite (Figure 5.4).

Using two different types of organically modified nanoclays (montmorillonite ion-exchanged with dimethyl-octadecylamine and synthetic fluoromica ion-exchanged with dimethyl di-tallow ammonium), the researchers observed an inversely proportional relationship between the nanocomposite organoclay content and the sample's *d*-spacing upon X-ray diffraction characterization (Figure 5.5). They attributed the phenomenon to the decrease in the available gallery space in the clay's silicate layers due to

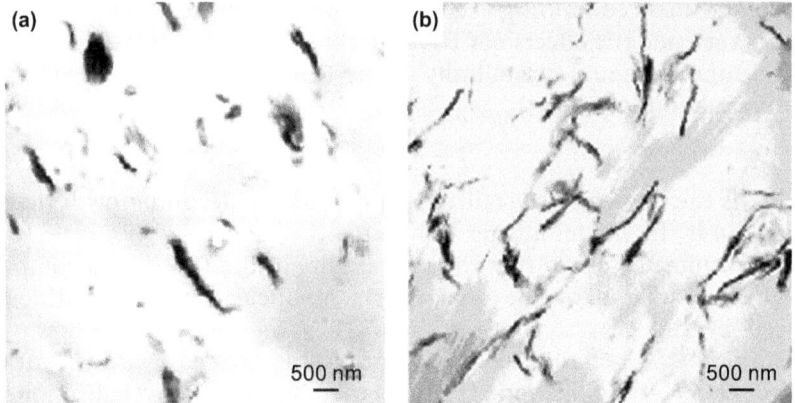

Figure 5.4 TEM of PHB/layered silicate nanocomposite: (a) PHBCN2.3 (2.3 wt% MMT), (b) PHBCN2 (2 wt% MAE).
Reprinted from Maiti *et al.*[38] with permission from the American Chemical Society.

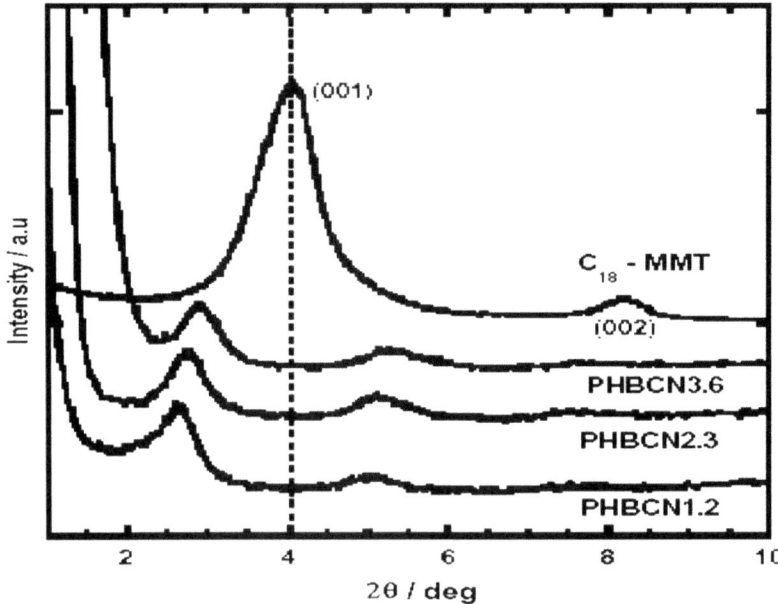

Figure 5.5 WAXD pattern of PHB/nanoclay composites.
Reprinted from Maiti *et al.*[38] with permission from the American Chemical Society.

the presence of a small amount of nanofiller, which in turns reflects the increase in the *d*-spacing to accommodate the intercalated polymer chains.[38]

Using melt intercalation techniques, Zhang *et al.*[34] prepared nanocomposites of poly 3-hydroxybutyrate-*co*-3-hydroxyhexanoate (PHBHHx)/layered

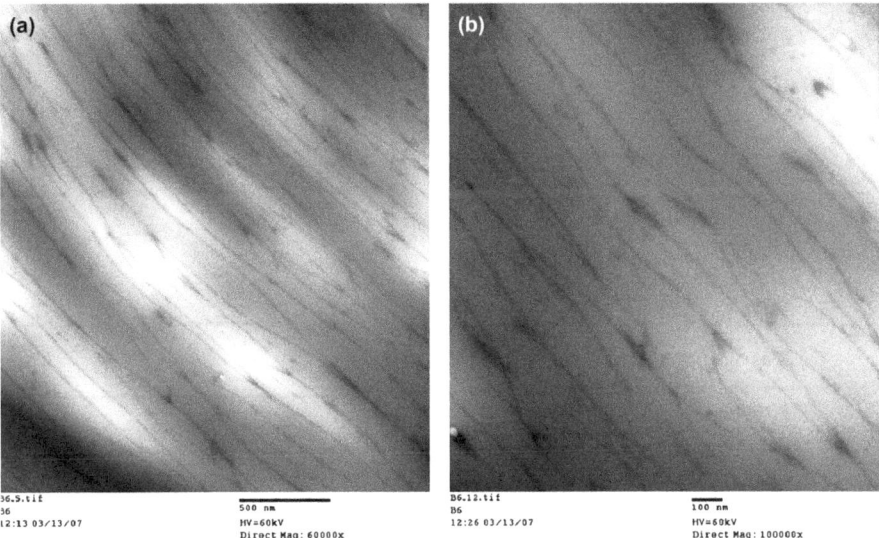

Figure 5.6 TEM images of PHBHHx nanocomposites: (a) PHBHHx/layered silicate (b) PHBHHx/expanded graphite; viewed at ×6000 magnification. Reprinted from Zhang *et al.*[34] with permission from Taylor & Francis.

silicates and PHBHHx/expanded graphite. For the nanofillers used in the two nanocomposites, at a lower nanofiller content, the researchers observed an exfoliated morphology with good dispersion (Figure 5.6).

Efficient dispersion of the nanofillers resulted in improved thermo–mechanical properties (Figure 5.7). In comparison to neat PHBHHx, the authors observed that the onset of the thermal degradation temperature decreases with an increase in dispersed nanofillers in both nanoclay (NC) and expanded graphite (NG) composites (Figure 5.7). In PHBHHx-nanoclay composites, the presence of 7 parts nanofiller to 100 parts polymer (NC7) resulted in about a 6.4% decrease in the thermal degradation temperature of the polymer (Figure 5.7(a)). This corresponds to about a five-fold decrease in polymeric toughness (Figure 5.7(c)). A decrease in the thermal degradation temperature of the polymer upon compositing with expanded graphite was also observed, but was not as prominent as in nanoclay composites. Compositing the neat polymer with 6 parts of NG in 100 parts of PHBHHx results in about a 3.9% reduction of the thermal degradation temperature (Figure 5.7(b)), incurring about a 2.4-fold reduction in toughness (Figure 5.7(c)). The effects of the two nanofiller composites are further highlighted in Figure 5.7(d), where compositing the neat polymer with 3 parts of nanoclay results in a 0.5% reduction of the polymer's elongation to break. Changing the nanofiller from nanoclay to 2 parts expanded graphite results in a considerable decrease in the elongation to break of the polymer (4%). Although melt intercalation proved to be an attractive process compared to solution casting, care has to be taken to avoid thermal degradation

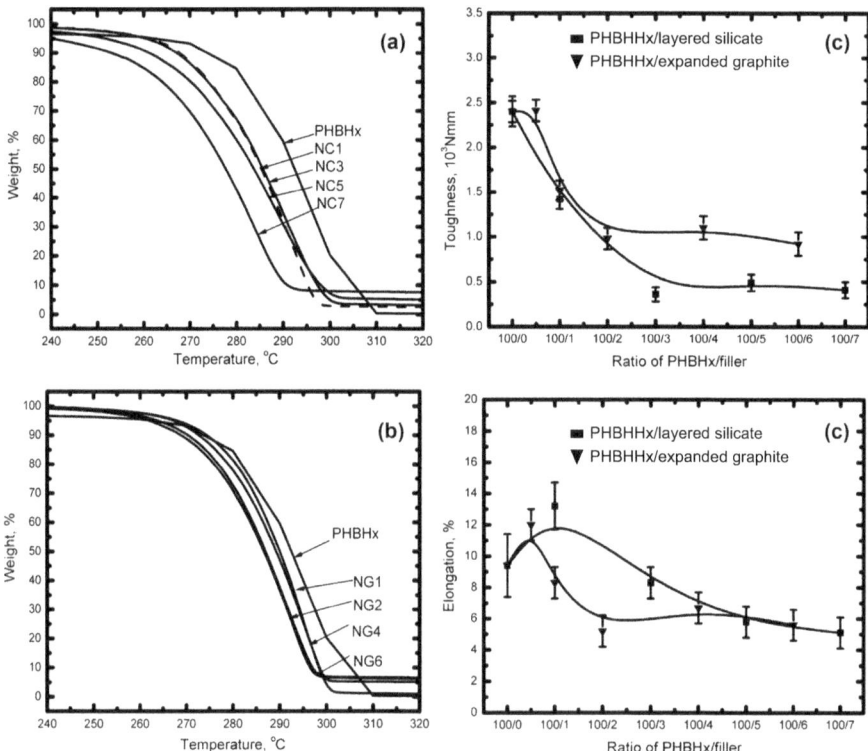

Figure 5.7 PHBHHx nanocomposites as a function of nanofiller content: (a) TGA thermogram of a PHBHHx/layered silicate, (b) TGA thermogram of PHBHHx/expanded graphite, (c) toughness, and (d) elongation to break. Reprinted from Zhang *et al.*[34] with permission from Taylor & Francis.

of the polymer during processing.[49] In addition, modified silicate quaternary ammonium cations were reported to greatly enhance PHB and PHBV degradation.[3,50,51] It is therefore crucial to optimize the processing parameters in order to limit the extent of degradation.

5.2.3 *In situ* Intercalative Polymerization

In this method, monomers are polymerized in between the layered nanofillers. The nanofillers are allowed to swell in a liquid monomer solution, followed by initiation of the polymerization using a suitable initiator such as heat, radiation, light, catalyst *etc.* resulting in nanocomposites of a linearly or crosslinked polymer matrix. It is said to be the first method used to synthesize polymer-layered silicate nanocomposites based on polyamide 6,[21] and is still widely employed in several nanocomposite fabrications such as in thermosetting polymer-layered silicate nanocomposites.[21] However, only a few studies have attempted to prepare a biodegradable polymer nanocomposite using this method.[52–55] Gorrasi *et al.*[56] obtained exfoliated nanocomposites of a poly-6-hydroxyhexanoates matrix through *in situ* intercalative ring opening polymerization of caprolactone with modified montmorillonite in the presence of a dibutyltin dimethoxide catalyst. Reports on the application of *in situ* intercalative polymerization for the production of bacterial PHA nanocomposites, especially those containing a medium-chain-length poly-3-hydroxyalkanoates matrix, are virtually non-existent.

The use of toxic organometallic catalysts and initiators in polymerization, as well as bulk solvents for monomer solutions, are among the industrial limitations of the *in situ* intercalative polymerization method, despite its reported advantage of good nanofiller dispersion in nanocomposite production.[3]

5.3 Characterization Methods for PHA Nanocomposites

When using layered nanofillers such as layered silicate, poor dispersion of nanofillers in the matrix results in aggregate (tactoids) nanocomposites. Interactions between the nanofillers and some of the extended polymer chains, however, result in an intercalated nanocomposite with an alternate arrangement of polymer and layered nanofiller (Figure 5.1). Absolute and homogeneous dispersion of nanofillers in the polymer matrix gives a nanocomposite with an exfoliated morphological characteristic (Figure 5.1). Previously, this ordered exfoliation was thought to be induced by steric interactions.[10,31] However, Zhang *et al.*[32] suggested that dispersion of the polymeric side chain into the nanofiller's layered spaces results in the exfoliation. The characterization of the state of the nanoparticles' dispersion allows the interpretation of the preceding morphologies, and the structural characterization relies heavily on techniques such as X-ray diffraction (XRD), wide angle X-ray diffraction (WAXD), simultaneous small angle X-ray scattering (SAXS) and electron microscopy (transmission, TEM or scanning, SEM). XRD is easy to use and is the most commonly employed method in establishing the nanocomposites' structural morphology, crystallinity as well as polymer melt intercalation.[9,33,57] For example, in intercalative polymer nanocomposites, the expansion of interlayer spacing (*d*-spacing) to allow polymer intercalation results in a shift of the X-ray diffraction peak towards lower values of 2θ (Figure 5.5). In each case, the *d*-spacing (Å) of the silicate layers can be easily calculated from the X-ray diffractogram 2θ values using Bragg's equation (eqn (5.2)),

$$d = \frac{\lambda}{2 \sin \theta} \qquad (5.2)$$

where λ is the X-ray wavelength. On the other hand, exfoliated nanocomposites exhibit excessive delamination of the silicate layer resulting in

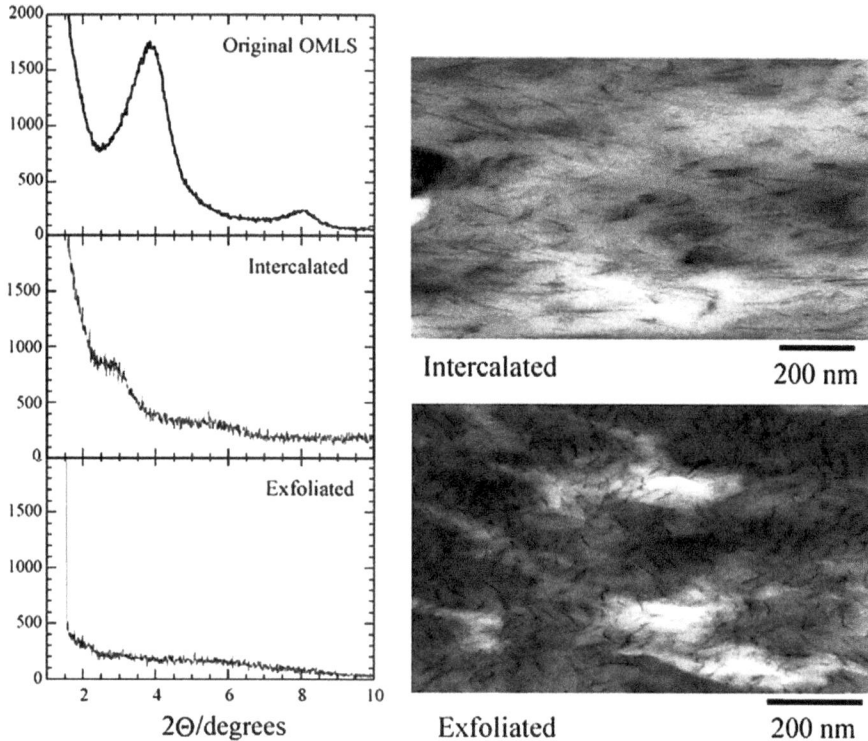

Figure 5.8 WAXD patterns and TEM images of different types of nanocomposites. Reprinted from Sinha Ray and Okamoto[33] with permission from Elsevier.

extensive layer separation, which results in the total disappearance of the X-ray diffraction peak of the crystalline nanoparticles (Figure 5.8).[9,33] Although TEM and SEM are time intensive processes, they are equally essential in revealing the nanocomposites' qualitative properties such as topology, structural architecture as well as defects in composite formation, nanoparticle distribution and dispersion within the polymer matrix (Figure 5.6).

Recently, Zhijiang *et al.*[58] employed the use of field emission scanning electron microscopy (FESEM) to qualitatively characterize a PHA/bacterial cellulose nanocomposite (Figure 5.9).

In addition to morphological characterization, analysis techniques such as FTIR[58] and NMR[33] were also reported to be employed in the structural characterization of the nanocomposites. For example, the ester carbonyl group of neat PHA usually shows a vibration at 1724–1731 cm^{-1}. In a nanocomposite, this peak is normally moved to a lower wavelength below 1724 cm^{-1} (Figure 5.10), probably due to reported hydrogen bond formation with the composite nanofiller.[36]

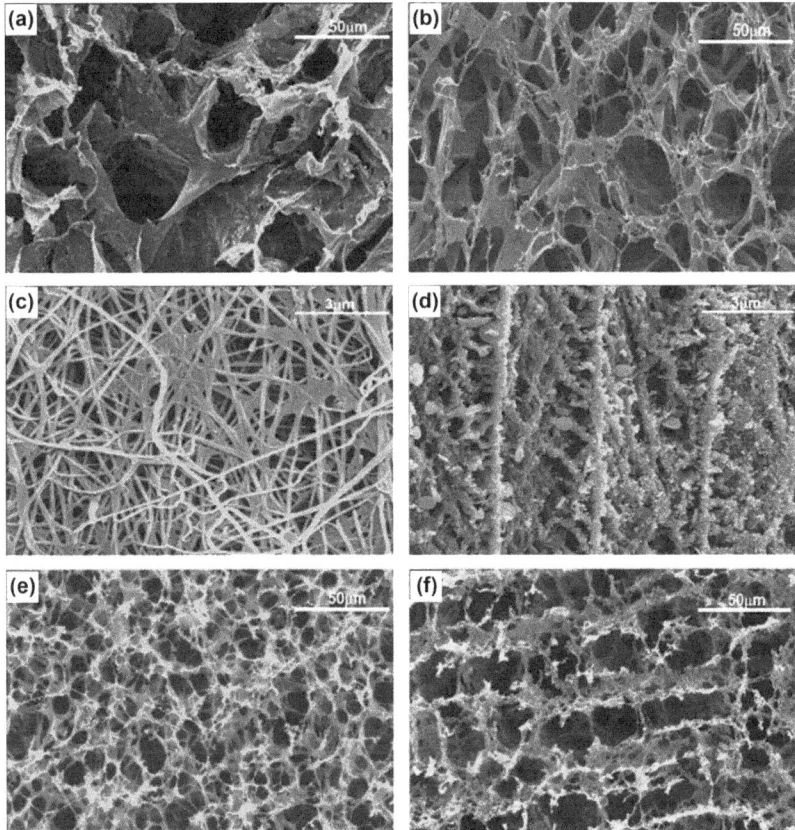

Figure 5.9 FESEM images of P(3HB-*co*-4HB) (a: surface morphology; b: cross-section morphology), BC (c: surface morphology; d: cross-section morphology) and P(3HB-*co*-4HB)/BC nanocomposite scaffolds (e: surface morphology; f: cross-section morphology).
Reprinted from Zhijiang *et al.*[58] with permission from Elsevier.

5.4 Applications of PHA Nanocomposites

As highlighted earlier, the incorporation of nanofillers into PHAs extends their application ranges while retaining their excellent biodegradability and compatibility. Recently, various types of biopolymer nanocomposites with properties suitable for a wide range of applications were prepared and characterized.[17,34,42,58,60] The poly(3-hydroxybutyrate-*co*-4-hydroxybutyrate)/ bacterial cellulose nanocomposite was used as a scaffold to efficiently cultivate Chinese hamster lung (CHL) fibroblast cells.[58] A nanocomposite based on poly-3-hydroxybutyrate and bioglass with high porosity was reported to be a potential candidate for bone tissue scaffolds (Table 5.2).[36] Smart drug delivery devices are normally composed of a targeting ligand that can direct the drug carrier to the specific site. In human cancer cells, a high expression

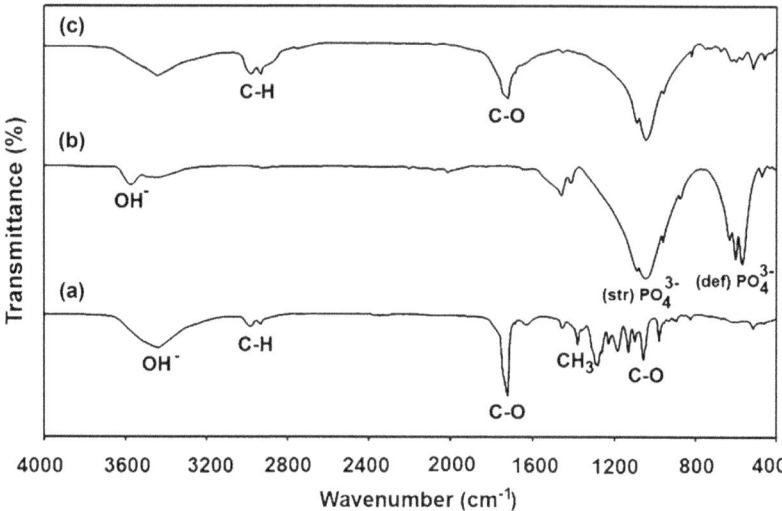

Figure 5.10 FT-IR absorption spectrum, (a) neat P3HB, (b) hydroxyapatite (HAp) scaffold and (c) HAp/P3HB scaffold.
Reprinted from Foroughi *et al.*[59] with permission from Springer.

of several receptors, such as those of folic acid,[61] and transferring,[62] was discovered using the ligand targeting strategy mentioned. Zhang *et al.*[63] reported the use of nanocomposite particles based on a PHA/folate ligand, which carries doxorubicin as an anticancer therapy (Table 5.2).

Some important industrial organic solvents (*e.g.* ethanol, 1,4-dioxane *etc.*) have excellent water-blending and miscibility features coupled with boiling points that are close to each other. Such azeotropic mixtures can be separated using conventional distillation processes in the presence of entrainers (*e.g.* benzene or cyclohexane), albeit with considerable difficulty.[64] During industrial wastewater treatment, achieving efficient separation of a mixture of azeotropes is a constant challenge to process designers and is usually addressed by a pervaporation process. Recently, a nanocomposite membrane based on a PHB-functionalized multi-walled carbon nanotubes/chitosan matrix (Table 5.2) was used to efficiently pervaporate a mixture of water and 1,4-dioxane without the use of an entrainer.[64] The researchers reported that the membrane showed an excellent selectivity with relatively high permeation flux towards water compared to existing commercial membranes.[64] Shape memory nanocomposite polymers have the ability to revert back to their original shape after being deformed thus expanding their applications to areas such as dry adhesion, microfluidics, biosensors, tissue engineering *etc.* (Table 5.2). In most cases, chain vitrification or crystallization as a result of inter-chain covalent association are said to be responsible for shape fixing in these kinds of polymeric materials.[65] Recently, Ishida *et al.*[65] reported the synthesis of a shape memory polymer nanocomposite derived

Table 5.2 Potential applications of PHA nanocomposites.

Polymer	Nanoparticle	Application	Material	Ref.
Poly(3-hydroxybutyrate-*co*-4-hydroxybutyrate)	Bacterial cellulose nanofibrils	Scaffold for CHL fibroblast cells		58
Poly(3-hydroxybutyrate)	Bioglass	Bone tissue scaffold		36
Poly(3-hydroxybutyrate-*co*-3-hydroxyoctanoate)	Doxorubicin/folic acid	Cancer drug carrier		63

Table 5.2 *(Continued)*

Polymer	Nanoparticle	Application	Material	Ref.
PHB	Multi-walled carbon nanotubes/ chitosan matrix	Azeotropic pervaporation		64
Poly(3-hydroxyoctanoate-*co*-3-hydroxyundecenoate)	Silsesquioxane	Biosensor		65
PHB	Silver sulfide	Catalyst		66

from poly(3-hydroxyoctanoate-*co*-3-hydroxyundecenoate) composited with nanofillers of silsesquioxane (POSS).

Silver ions are thought to adhere to the negatively charged bacterial cell wall, changing the permeability of the cell wall and consequently causing protein denaturation, which induces cell lysis and death. This specific trait of silver nanoparticles is exploited in food packaging and processing industries. Nanocomposites of biodegradable polymers containing silver nanoparticles exhibited antimicrobial activities, which inhibit the growth of post-processing microbial contaminants, thereby extending the shelf life of food and improving food safety.[60] In fact, such metal intercalated nanoparticles not only posses antimicrobial properties, but are also reported to exhibit catalytic, electrical and sensing properties (Table 5.2).[66]

The current high demand for portable electronic gadgets warrants the urgent need of a portable power source to support these appliances. At present, lithium-ion batteries use a liquid electrolyte that is flammable, and can cause problems such as gas formation due to overcharging, and release of toxic and hazardous materials into the environment following end-of-cycle disposal.[67] An alternative to this liquid electrolyte in electrochemical cells is a system consisting of a solid polymer electrolyte produced through incorporation of an electrochemical salt (*e.g.* lithium salt) in the polymer matrix. This kind of polymeric cell is expected to show a better charge cycle than the conventional liquid electrolyte systems in existence today.[68] In this respect, Reusch and Reusch[69] described an embodiment for the method of preparing such polymer electrolyte cells using polyhydroxyalkanoate nanocomposites.[69]

5.5 Conclusions

The increasing use of non-degradable fossil-derived plastics has caused great concern in terms of environmental pollution and degradation issues. Biodegradable polymers such as bacterial polyhydroxyalkanoates (PHAs) are considered as an alternative to existing petrochemical-based plastics. Despite their promising commercial potential, most of the PHAs produced exhibit brittleness, a low heat distortion temperature and poor gas-barrier properties, thus limiting their applications in vast potential fields. Incorporation of nanofillers into the biodegradable polymer matrices resulted in nanocomposites that possess significantly improved mechanical and thermal properties both in solid and melt states along with enhanced gas barrier properties (O_2, CO_2, water vapor *etc.*) compared to conventional filler composites. These superior qualities extend the polymeric composites' applications to aggressive environments where the neat polymers are likely to fail.

However, extensive research on the scarcely explored field of nanocomposite formulation based on polyhydroxyalkanoates is needed in order to enhance their mechanical, and other, properties, which can be manipulated to suit end-users' requirements in order to meet a wide range of applications.

References

1. M. S. M. Annuar, I. K. P. Tan, S. Ibrahim and K. B. Ramachandran, *Food Bioprod. Process.*, 2007, **85**, 104–119.
2. K. Sudesh, *Polyhydroxyalkanoates from Palm Oil: Biodegradable Plastics*, Springer, Germany, 2013, pp. 63–77.
3. V. Ojijo and S. Ray, *Prog. Polym. Sci.*, 2013, **38**(10–11), 1543–1589.
4. A. M. Gumel, M. S. M. Annuar and T. Heidelberg, *PLoS One*, 2012, 7, e45214.
5. A. M. Gumel, M. S. M. Annuar and Y. Chisti, *J. Polym. Environ.*, 2013, **21**, 580–605.
6. A. Poli, P. Di Donato, G. R. Abbamondi and B. Nicolaus, *Archaea*, 2011, **2011**, DOI: 10.1155/2011/693253.
7. S. Ray and M. Bousmina, *Prog. Mater. Sci.*, 2005, **50**, 962–1079.
8. C. Johansson, in *Nanocomposites with Biodegradable Polymers: Synthesis, Properties and Future Perspectives*, ed. V. Mittal, Oxford University Press, Oxford, 2011, pp. 348–367.
9. S. Ray and M. Bousmina, *Biodegradable polymer/layered silicate nano-composites*, WoodHead Publishing, Cambridge, England, 2006.
10. J. K. Pandey, K. Raghunatha Reddy, A. Pratheep Kumar and R. Singh, *Polym. Degrad. Stab.*, 2005, **88**, 234–250.
11. Q. Zhang, Q. Liu, J. E. Mark and I. Noda, *Appl. Clay Sci.*, 2009, **46**, 51–56.
12. S. K. Misra, F. Ohashi, S. P. Valappil, J. C. Knowles, I. Roy, S. R. P. Silva, V. Salih and A. R. Boccaccini, *Acta Biomater.*, 2010, **6**, 735–742.
13. S. K. Misra, P. Watts, S. P. Valappil, S. Silva, I. Roy and A. Boccaccini, *Nanotechnology*, 2007, **18**, 075701.
14. S. Wang, C. Song, G. Chen, T. Guo, J. Liu, B. Zhang and S. Takeuchi, *Polym. Degrad. Stab.*, 2005, **87**, 69–76.
15. P. Bordes, E. Pollet and L. Avérous, *Prog. Polym. Sci.*, 2009, **34**, 125–155.
16. H.-y. Yu, Z.-y. Qin and Z. Zhou, *Progress in Natural Science: Materials International*, 2011, **21**, 478–484.
17. K. L. Dagnon, H. H. Chen, L. H. Innocentini-Mei and N. A. D'Souza, *Polym. Int.*, 2009, **58**, 133–141.
18. R. Zhang, H. Huang, W. Yang, X. Xiao and Y. Hu, *Composites, Part A*, 2012, **43**, 547–552.
19. H. Ishida, S. Campbell and J. Blackwell, *Chem. Mater.*, 2000, **12**, 1260–1267.
20. H. Träubel, in *New Materials Permeable to Water Vapor*, Springer, Berlin Heidelberg, 1999, pp. 133–152.
21. Z. Shen, G. P. Simon and Y.-B. Cheng, *Polymer*, 2002, **43**, 4251–4260.
22. Z. Zhang, L. Zhang, Y. Li and H. Xu, *Polymer*, 2005, **46**, 129–136.
23. C.-L. Chiang and C.-C. M. Ma, *Eur. Polym. J.*, 2002, **38**, 2219–2224.
24. J.-A. He, R. Valluzzi, K. Yang, T. Dolukhanyan, C. Sung, J. Kumar, S. K. Tripathy, L. Samuelson, L. Balogh and D. A. Tomalia, *Chem. Mater.*, 1999, **11**, 3268–3274.
25. J. Musil and P. Baroch, *Vacuum*, 2013, **87**, 96–102.

26. Y. Leconte, S. Veintemillas-Verdaguer, M. Morales, R. Costo, I. Rodríguez, P. Bonville, B. Bouchet-Fabre and N. Herlin-Boime, *J. Colloid Interface Sci.*, 2007, **313**, 511–518.

27. R. A. McGill, D. B. Chrisey and A. Pique, U. S. Patent, United States, US 6660343 B2, 2003.

28. E. Lee, D. Mielewski and R. Baird, *Polym. Eng. Sci.*, 2004, **44**, 1773–1782.

29. D. L. Tomasko, X. Han, D. Liu and W. Gao, *Curr. Opin. Solid State Mater. Sci*, 2003, **7**, 407–412.

30. Y. Lin, X. Cui, C. Yen and C. M. Wai, *J. Phys. Chem. B*, 2005, **109**, 14410–14415.

31. C. M. Koo, H. T. Ham, S. O. Kim, K. H. Wang, I. J. Chung, D.-C. Kim and W.-C. Zin, *Macromolecules*, 2002, **35**, 5116–5122.

32. X. Zhang, G. Lin, R. Abou-Hussein, M. K. Hassan, I. Noda and J. E. Mark, *Eur. Polym. J.*, 2007, **43**, 3128–3135.

33. S. Sinha Ray and M. Okamoto, *Prog. Polym. Sci.*, 2003, **28**, 1539–1641.

34. X. Zhang, G. Lin, R. Abou-Hussein, W. M. Allen, I. Noda and J. E. Mark, *J. Macromol. Sci., Part A*, 2008, **45**, 431–439.

35. S. Bruzaud and A. Bourmaud, *Polym. Test.*, 2007, **26**, 652–659.

36. H. Hajiali, S. Karbasi, M. Hosseinalipour and H. R. Rezaie, *J. Mater. Sci.: Mater. Med.*, 2010, **21**, 2125–2132.

37. P. Bordes, E. Hablot, E. Pollet and L. Avérous, *Polym. Degrad. Stab.*, 2009, **94**, 789–796.

38. P. Maiti, C. A. Batt and E. P. Giannelis, *Biomacromolecules*, 2007, **8**, 3393–3400.

39. A. Botana, M. Mollo, P. Eisenberg and R. M. Torres Sanchez, *Appl. Clay Sci.*, 2010, **47**, 263–270.

40. R. A. Vaia and E. P. Giannelis, *Macromolecules*, 1997, **30**, 7990–7999.

41. B. K. G. Theng, *Formation and properties of clay-polymer complexes*, access online *via* Elsevier, 2012.

42. C. Zhijiang, Y. Guang and J. Kim, *Curr. Appl. Phys.*, 2011, **11**, 247–249.

43. E. Ten, J. Turtle, D. Bahr, L. Jiang and M. Wolcott, *Polymer*, 2010, **51**, 2652–2660.

44. E. Ten, D. F. Bahr, B. Li, L. Jiang and M. P. Wolcott, *Ind. Eng. Chem. Res.*, 2012, **51**, 2941–2951.

45. J. Yang, B. J. McCoy and G. Madras, *J. Chem. Phys.*, 2005, **122**, 064901.

46. M. Avrami, *J. Chem. Phys.*, 1939, 7, 1103.

47. R. A. Vaia, H. Ishii and E. P. Giannelis, *Chem. Mater.*, 1993, **5**, 1694–1696.

48. W. Mook Choi, T. Wan Kim, O. Ok Park, Y. Keun Chang and J. Woo Lee, *J. Appl. Polym. Sci.*, 2003, **90**, 525–529.

49. M. C. Sin, M. S. M. Annuar, I. K. P. Tan and S. N. Gan, *Polym. Degrad. Stab*, 2010, **95**, 2334–2342.

50. P. Bordes, E. Pollet, S. Bourbigot and L. Avérous, *Macromol. Chem. Phys.*, 2008, **209**, 1473–1484.

51. E. Hablot, P. Bordes, E. Pollet and L. Avérous, *Polym. Degrad. Stab.*, 2008, **93**, 413–421.

52. V. Katiyar, N. Gerds, C. B. Koch, J. Risbo, H. C. B. Hansen and D. Plackett, *Polym. Degrad. Stab.*, 2010, **95**, 2563–2573.
53. J.-M. Raquez, Y. Habibi, M. Murariu and P. Dubois, *Prog. Polym. Sci.*, 2013, **38**(10), 1504–1542.
54. L. Urbanczyk, F. Ngoundjo, M. Alexandre, C. Jérôme, C. Detrembleur and C. Calberg, *Eur. Polym. J.*, 2009, **45**, 643–648.
55. G. Gorrasi, M. Tortora, V. Vittoria, G. Galli and E. Chiellini, *J. Polym. Sci. Part B: Polym. Phys.*, 2002, **40**, 1118–1124.
56. G. Gorrasi, M. Tortora, V. Vittoria, E. Pollet, B. Lepoittevin, M. Alexandre and P. Dubois, *Polymer*, 2003, **44**, 2271–2279.
57. P. C. LeBaron, Z. Wang and T. J. Pinnavaia, *Appl. Clay Sci.*, 1999, **15**, 11–29.
58. C. Zhijiang, H. Chengwei and Y. Guang, *Carbohydr. Polym.*, 2012, **87**, 1073–1080.
59. M. R. Foroughi, S. Karbasi and R. Ebrahimi-Kahrizsangi, *J. Porous Mater.*, 2012, **19**, 667–675.
60. J.-W. Rhim, H.-M. Park and C.-S. Ha, *Prog. Polym. Sci.*, 2013, **38**(10), 1629–1652.
61. S.-H. Kim, J.-K. Kim, S.-J. Lim, J.-S. Park, M.-K. Lee and C.-K. Kim, *Eur. J. Pharm. Biopharm.*, 2008, **68**, 618–625.
62. T. Kakudo, S. Chaki, S. Futaki, I. Nakase, K. Akaji, T. Kawakami, K. Maruyama, H. Kamiya and H. Harashima, *Biochemistry*, 2004, **43**, 5618–5628.
63. C. Zhang, L. Zhao, Y. Dong, X. Zhang, J. Lin and Z. Chen, *Eur. J. Pharm. Biopharm.*, 2010, **76**, 10–16.
64. Y. T. Ong, A. L. Ahmad, S. H. S. Zein, K. Sudesh and S. H. Tan, *Sep. Purif. Technol.*, 2011, **76**, 419–427.
65. K. Ishida, R. Hortensius, X. Luo and P. T. Mather, *J. Polym. Sci. Part B: Polym. Phys.*, 2012, **50**, 387–393.
66. S. Yeo, W. Tan, M. Abu Bakar and J. Ismail, *Polym. Degrad. Stab.*, 2010, **95**, 1299–1304.
67. C. Polo Fonseca and S. Neves, *J. Power Sources*, 2006, **159**, 712–716.
68. C. P. Fonseca, D. S. Rosa, F. Gaboardi and S. Neves, *J. Power Sources*, 2006, **155**, 381–384.
69. R. N. Reusch and W. H. Reusch, in Google, U. patents, Google Patents, United States, 1993, vol. US5266422.

CHAPTER 6

Polyhydroxyalkanoate-based Multiphase Materials

DHRITI KHANDAL, ERIC POLLET AND LUC AVÉROUS*

BioTeam/ICPEES-ECPM, UMR 7515, Université de Strasbourg,
25 rue Becquerel, 67087 Strasbourg Cedex 2, France
*Email: luc.averous@unistra.fr

6.1   Introduction

A growing demand for bio-based plastics has been gaining impetus over the past few decades simply out of a need to be less dependent on non-renewable resources and to address growing environmental concerns. Bio-based plastics offer several advantages over their petroleum-based counterparts with regard to their cost, biodegradability, and their sources are not confined to any particular country. Apart from this, synthetic polymers derived from petroleum-based monomers have been a huge risk to nature and wildlife, *e.g.*, disposable plastics bags have been known to endanger aquatic life.[1] The applications of polymers in different domains of life, such as packaging, drug delivery, electrical appliances and commodities, require the possibility of varying not only their mechanical properties but also the duration of, and the toxicity resulting from, biodegradation. Bio-based plastics can be modified by traditional means such as extrusion and injection moulding, their mechanical properties can be varied by physical blending or chemical modification, and their susceptibility to biodegradation is subject to the degree of chemical modification.[2] The poor degradation rate of fossil-based plastics requires valorisation of their waste, which presents environmental issues as well. For example, their incineration

RSC Green Chemistry No. 30
Polyhydroxyalkanoate (PHA) Based Blends, Composites and Nanocomposites
Edited by Ipsita Roy and Visakh P M
© The Royal Society of Chemistry 2015
Published by the Royal Society of Chemistry, www.rsc.org

for energy production yields toxic emissions (*e.g.*, dioxin) while material valorisation is limited due to the difficulties in validating an economically relevant method. In addition, plastic recycling shows a negative eco-balance due to the necessity of large volumes of water and energy consumption (waste grinding and plastic processing) during all the different phases of recycling. As plastics represent a large part of the waste collection at local, regional, and national levels, institutions are now aware of the significant savings that compostable or biodegradable materials would generate. Thus, for these different reasons, it has become important to replace conventional plastics with biodegradable polymers, particularly for packaging applications that are of major interest to society (industry to citizens to nature).

The potential of biodegradable and bio-based polymers is being recognized as seen in growing research areas involved in helping us to overcome the limitations of petrochemical resources at present and in the future. Fossil fuels and natural gas can be partially replaced by greener agricultural sources, which would also participate in the reduction of CO_2 emissions. The major drawback in preventing their large scale exploitation is the high production cost of bacterial fermentation-based polymers, the poor mechanical properties of the agricultural-based polymers, and industrially difficult methods of homogeneous modifications. However, biodegradable and bio-based polymers often present macromolecular architectures and so new properties that cannot be found with conventional polymers.

6.1.1   Renewability and Sustainable Development

The concept of renewability and sustainable development is often described and understood in terms of "carbon economy". The UN World Commission on "Environment and Development in our Future" defines sustainability as the development that meets the needs of the present time without compromising on the ability of future generations to meet their own needs. One approach that meets this requirement is the "concept of reincarnation" or the "cradle to grave" approach, which pertains to short term applications of polymers such as packaging.[1] The use of annually renewable biomass must be understood in terms of carbon economy or carbon cycle. A low carbon economy (LCE) refers to the minimum greenhouse gas emissions in the atmosphere, specifically CO_2 and methane. The common approaches to achieving LCE are using renewable sources and technology that is energy efficient. The carbon cycle, on the other hand, refers to the source of carbon and the final destination of the converted carbon. The four main carbon sources on the planet are: the lithosphere (*e.g.* limestone), the biosphere (vegetal and animal), the hydrosphere (*e.g.* bicarbonate dissolved in the oceans) and the atmosphere (CO_2, CH_4); the carbon cycle is thus the complex process in which carbon can get exchanged between the four main reservoirs of carbon. The exchange of carbon among these four reservoirs remains balanced by natural processes but recent human activities (burning fossil fuel and massive deforestation) have led to an important imbalance in the carbon economy and carbon cycle. The human activities have resulted in a huge and rapid release of CO_2 (and other greenhouse effect gases) in the atmosphere which cannot be fully compensated by photosynthesis activity and dissolution in the oceans. It results in a large accumulation of carbon dioxide in the atmosphere; this greenhouse gas, along with CH_4 (also referred to as GHG), is contributing to global warming. People are now aware that efforts have to be made to re-balance the carbon cycle by reducing the amount of CO_2 in the atmosphere. Part of the carbon cycle re-balancing concept is based on the development and manufacture of products based on renewable and biodegradable resources. By collecting and composting biodegradable plastic waste, we can generate much-needed carbon-rich compost: humus materials. These valuable soil inputs can go back to the farmland and ''reinitiate'' the carbon cycle. Then, the plants' growth contributes to reducing CO_2 atmospheric accumulation through photosynthesis. Besides, composting is an increasing key point to maintain the sustainability of the agricultural system by reducing the consumption of chemical fertilizers.

6.1.2 Biodegradability and ''Compostability''

According to ASTM standard D-5488-94d, a biodegradable article is capable of undergoing decomposition into carbon dioxide, methane, water, inorganic compounds, or biomass by enzymatic degradation in microbes. ''Compostability'' is similar to biodegradation and, as per ASTM, requires the plastic to break down into biomass, CO_2, and water at the same rate as water. Hence, biodegradation is the degradation of an organic material caused by biological activity, mainly micro-organisms' enzymatic action. This leads to a significant change in the material chemical structure. The end-products are carbon dioxide, new biomass and water (in the presence of oxygen: aerobic conditions) or methane (when oxygen is absent: anaerobic), as defined in the European Standard EN 13432:2000. There are different media (liquid, inert, or compost medium) to carry out and analyse biodegradability. Depending on the type of standard to be followed (ASTM, EN), different composting conditions (humidity, temperature cycle) must be realized to determine the compostability level.[3] Therefore, a comparison of the results obtained from different standards seems to be difficult or impossible. It is important henceforth, that the conditions of humidity and temperature be specified before mentioning the duration required for the biodegradation of the product.

Another aspect that we must also take into account is the amount of mineralization as well as the nature of the residues (commonly called ''by-products'') left after biodegradation.[4] The accumulation of contaminants with toxic residues in the compost can cause plant growth inhibition in these products, which must serve as fertilizers. The key issue is to determine the environmental toxicity level for these by-products, which is known as eco-toxicity.[5] Even if the biodegradation behaviour of a polymer is difficult to predict, some general rules enable the determination of the biodegradability evolution. For example, an increase in parameters such as the hydrophobicity, the macromolecules' molar masses, and the crystallinity or the size of spherulites decreases the biodegradability.[6]

6.1.3 Biodegradable Polymer Classifications

Biodegradable polymers are a growing field of investigation.[7–9] A vast number of biodegradable polymers (*e.g.* cellulose, chitin, starch, polyhydroxyalkanoates, polylactide, polycaprolactone, collagen and other polypeptides) are formed in nature during the growth cycles of all organisms or are synthesized. Some micro-organisms and enzymes capable of degrading such polymers have been identified.[7,10] Different classifications of various biodegradable polymers have been proposed in Figure 6.1. As an example, we can propose to classify the biodegradable polymers according to their synthesis process: (i) polymers from biomass such as the agro-polymers obtained from agro-resources (*e.g.* starch or cellulose), (ii) polymers obtained by microbial fermentation such as the polyhydroxyalkanoates (PHAs), (iii) polymers conventionally and chemically synthesized using monomers obtained from agro-resources, *e.g.* polylactic acid (PLA), (iv) polymers obtained by chemical synthesis from petroleum-based monomers such as polystyrene (PS). More and more, bio-based polymers are developed as replacements for this last category of non-renewable polymers, such as bio-based aliphatic and aromatic copolyesters, *e.g.* renewable PBSA or PBS.

6.1.4 Polyhydroxyalkanoates' Classification and Chemical Structures

Polyhydroxyalkanoates (PHAs) are a family of biopolymers synthesized by many bacteria as intracellular carbon and energy storage granules. Since PHAs are mainly produced from renewable resources by fermentation, PHAs are in agreement with the rising worldwide concept of sustainable development and are classified as environmentally friendly materials. Besides, PHAs are considered as biodegradable, thus are suitable for *e.g.* short term packaging, and also as biocompatible in contact with living tissues, which makes them candidates for biomedical applications (*e.g.* drug encapsulation, tissue engineering). PHAs can be degraded by abiotic degradation *i.e.* simple hydrolysis of the ester bond without requiring the presence of enzymes to catalyze this hydrolysis. During the biodegradation process, the enzymes degrade the residual products until final mineralization (biotic degradation).

Polyhydroxyalkanoates are generally classified into short-chain-length PHA (sCL-PHA) and medium-chain-length PHA (mCL-PHA) by the different number of carbons in their repeating units. For instance, sCL-PHAs contain 4 or 5 carbons in their repeating units, while mCL-PHAs contain 6 or more

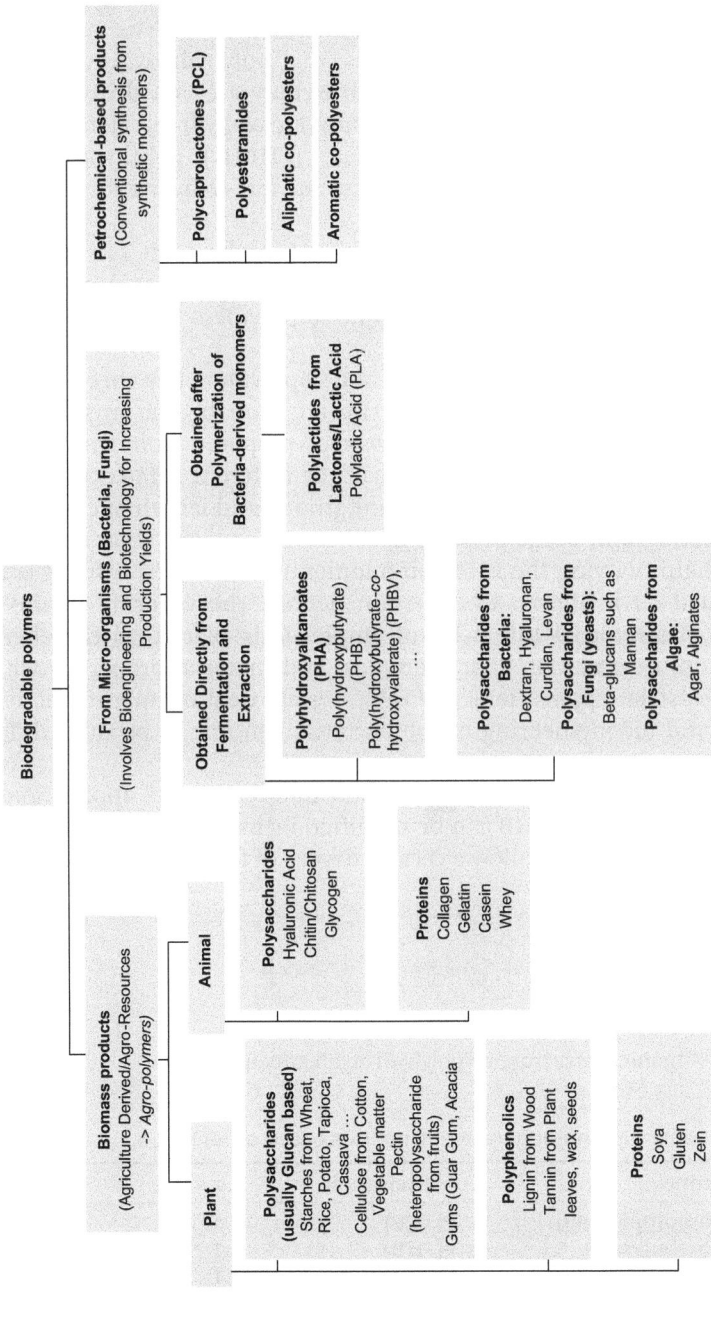

Figure 6.1 Classification of the main biodegradable polymers.

carbons in the repeating units. The term mCL was coined because the number of carbons in the monomers roughly corresponds to those of medium-chain-length carboxylic acids. PHA nomenclature may still be in a state of flux as new structures continue to be discovered. The main polymer of the polyhydroxyalkanoates family is the polyhydroxybutyrate homo-polymer (PHB), but different poly(hydroxybutyrate-*co*-hydroxyalkanoates) copolyesters exist, such as poly(hydroxybutyrate-*co*-hydroxyvalerate) (PHBV), or poly(hydroxybutyrate-*co*-hydroxyhexanoate) (PHBHx), poly(hydroxybutyrate-*co*-hydroxyoctanoate) (PHBO) or poly(hydroxybutyrate-*co*-hydroxyoctadecanoate) (PHBOd) (Figure 6.2 and Table 6.1).

Figure 6.2 shows the generic formula for PHAs where x is 1 or higher, and R can be either hydrogen or hydrocarbon chains of up to around C16 in length. The main members of the PHA homopolymers family are presented in Table 6.1.

A wide range of PHA homopolymers and copolymers have been produced, in most cases at the laboratory scale. PHAs are probably the largest family of thermoplastics with over 125 different types prepared from various hydro-xyalkanoate monomers. The properties of the different PHAs depend on the comonomer structure (length of pendant groups and length of carbon chain in the repeating unit of the polymer backbone) along with their mol% in the polymer chain. Varying the PHA comonomer and its mol% content is quite well pursued as it allows, to a certain degree, the tailoring of the PHA properties. Therefore, while a lot of emphasis is placed on the chemical modification, degradation characteristics, and material properties with re-gard to industrial applications of PHAs, a great deal of importance is also placed on the bioengineering of the bacterial genome to produce different types of PHAs.

PHB is a highly crystalline polymer and due to this, its applications are quite limited. However, PHB can be modified by incorporating comonomers in the polymer chain that allow a certain degree of flexibility in the chain and

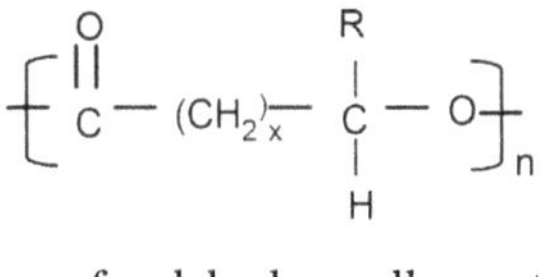

Figure 6.2 Chemical structure of polyhydroxyalkanoates.

Table 6.1 Some PHA homopolymer structures based on Figure 6.2.

Chemical name	Abbreviation	x value	R group
Poly(3-hydroxypropionate)	P(3HP)	1	Hydrogen
Poly(3-hydroxybutyrate)	P(3HB)	1	Methyl
Poly(3-hydroxyvalerate)	P(3HV)	1	Ethyl
Poly(4-hydroxybutyrate)	P(4HB)	2	Hydrogen
Poly(5-hydroxybutyrate)	P(5HB)	2	Methyl
Poly(5-hydroxyvalerate)	P(5HV)	3	Hydrogen

thereby reduce the crystalline character to a certain extent. For example, copolymerization with hydroxyvalerate, by the manipulation of the feedstock and bioprocessing conditions, affords a PHBV polymer that has comparatively more elastomeric properties and is commercially more viable than PHB.[11] Among the numerous PHA copolymers, only a few of them have attracted industrial interest and are commercialised.

The changes in the melting point and T_g, apart from improved elasticity and modulus properties of the PHA copolymers, allow the use of milder processing conditions (for example lowered melt temperatures for the processing). The copolymers of PHAs are typically random in sequence and obtained from mixed culture production.[12] Thus, poly(3-hydroxybutyrate-*co*-3-hydroxyvalerate) or P(3HB-*co*-3HV), also called PHBV, is based on a random arrangement of two monomers with R = methyl and R = ethyl and poly(3-hydroxybutyrate-*co*-3-hydroxyhexanoate) consists of two monomers with R = methyl and propyl. The mixed culture production of PHA involves the use of complex waste feedstocks such as sugar cane molasses, paper mill effluents, fermented municipal sludge, fermented olive oil mill effluents, industrial and domestic waste waters and so on.[13] The incorporation of different monomers in the polymer chain by the polymerizing enzyme PHA synthase is often regulated by genetic and metabolic engineering of the microbial strains.[14] It is, however, reported that when the substrate is an odd-C then the PHA polymer contains only odd-C monomers and when the substrate is an even-C then the PHA polymer chains have only even-C monomers. Unsaturated monomers have also been successfully incorporated by using a 1-alkene carbon source.[15]

Mixed culture production not only aids in lowering the cost by the use of easily available cheap feedstocks but also provides an environmentally friendly way of managing waste. Another advantage of mixed culture production is that the life cycle inventories (LCI), *i.e.* the energy consumption and carbon dioxide emissions from "cradle to reincarnation", are more favourable for the PHAs random copolymers compared to pure PHB. Akiyama *et al.* have reported that even though the production costs of PHBHx and PHB were comparable, the energy consumption and carbon dioxide production favor the former over the latter.[16]

While bioengineering has led to numerous types of PHAs with varying properties, their crystallization kinetics after processing is also important in determining the final properties of the material. Thus, the crystalline forms and the need of nucleating agents in reducing the crystallization of the PHA copolymers are also important and have been briefly discussed in the following sections.

Though not completely resistant to UV radiation, being a linear aliphatic polyester, PHA has been reported to have better resistance to UV radiation than polypropylene.[17] Due to this property, a PHA (reported as a PHA-I6001 from Metabolix with a crystallinity of 25%) was blended with PVC (a UV sensitive polymer) to provide this latter higher UV resistance without promoting biodegradation. PHA, being miscible with PVC, did not lead to phase separation in the blend and was found to offset UV-induced yellowing of PVC providing colour hold throughout the entire UV exposure.[18]

6.1.5 Crystalline Structures of Polyhydroxyalkanoates

PHAs are optically active biological polyesters with the hydroxyalkanoic acid monomers having an R configuration due to the stereospecificity of the polymerizing enzyme. Once extracted from the cell, PHB presents high crystallinity. In contrast, *in vivo* PHB exists as an amorphous polymer.[14] It was initially thought that the amorphous state of the PHAs *in vivo* may be due to the presence of plasticizers or nucleation inhibitors within the cell inclusion. However, amorphous PHA was also produced *in vitro* showing that the process of crystallization of PHA upon its extraction may be the result of coalescing of the polymer chains that in turn accelerates the crystallization process.[19]

Though crystallinity of PHAs has been well documented in the literature along with their X-ray diffraction analysis, here we shall only briefly discuss certain salient features worth mentioning of the crystalline forms of PHB and some other PHAs. PHB can result in lamellar crystalline structures when crystallized from dilute solutions or into spherulites when prepared from melt processing. Typically, the crystals of P(3HB) are lath-shaped of 5–10 μm length with a thickness of 4–10 nm.[20]

PHB can take up the helix conformation (called the α-form) or a planar zig-zag conformation (called the β-form). The α-conformation results in lamellar crystals while the less common β-form is usually formed by extending the uniaxially oriented films of the PHB further, and occupies the amorphous regions between the lamellar crystals of the α-form.[21]

Zhang *et al.* tried to schematically represent the crystallization of PHA into the α-form and the β-form with respect to the different elaboration conditions (reproduced in Figure 6.3). As in the representation in Figure 6.3, and as suggested in various literature reports, the β-form occupies the amorphous region between the lamellar α-form.[22] It should be noted here that while both the β-form and the α-form can be found in the PHB films, it is the α-form that is more stable.[20]

The mCL-PHAs obtained from mixed cultures have also been reported to result in helix conformations forming an orthorhombic lattice with two molecules per unit cell. The side chains in the comonomers are believed to form ordered sheets, such that the longer the side chain, the thicker the lamellar sheet. Longer side chains are believed to form stable ordered sheets.

The crystallinity of copolymers is also affected by the mol% of the comonomer. The group of Ozaki evaluated the crystal structures in PHBV with regard to the intermolecular interactions and reported that for HV content less that 30%, the strength of the H-bonds between the methyl group and the carbonyl group of the HB part was similar to that of PHB. As the HV content in the copolymer increases, H-bonding begins to take place between

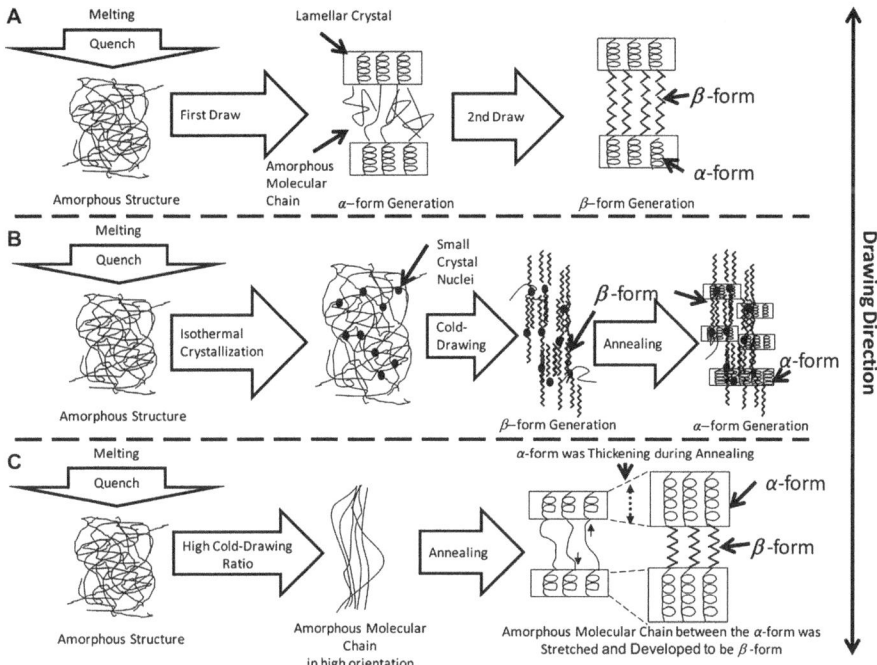

Figure 6.3 Schematic model of planar zig-zag conformation (β-form) generation. Model A: the β-form is produced in the amorphous region between lamellar crystals by two-step-drawing. Model B: the β-form is generated between small crystal nuclei which grow while samples are kept near T_g. Model C: the β-form is generated during the annealing process by thickening of the lamellar crystals (reproduced with permission from reference 22).

the methylene group and the carbonyl group of the HV groups.[23] They reported that the transition of the crystal structure from a PHB-type crystal to a PHV-type crystal begins to take place at around 50% HV content. Similar behaviour has also been observed for other copolymers such as P(3HB-*co*-HP) where the crystal structure changes when a certain concentration of the comonomer is reached. For the copolymer P(3HB-*co*-Hx), however, it has also been reported that the crystal structure is only slightly affected by the Hx content.[20]

6.2 PHA-based Multiphase Materials

6.2.1 Generalities

Extrusion of PHA-based materials is in general linked with another processing step such as thermoforming, injection moulding, fibre drawing, film blowing, bottle blowing, or extrusion coating. The properties of the polymer will therefore depend on the specific conditions during the processing steps *e.g.* the thermo–mechanical input. The main parameters during melt processing will be temperature, residence time, moisture content and atmosphere. But, the major problem in the manufacturing of PHA-based products is their high thermal sensitivity. PHA shows a low degradation temperature compared to its melting temperature. For instance, PHB homopolymer presents a narrow window for processing conditions. The PHB thermal[24–31] and thermo–mechanical[32,33] stabilities have been well described in the literature, demonstrating that the thermal degradation occurs according to a one-step process, namely a random chain scission reaction. Under extrusion, increasing the shear level, the temperature, and/or the residential time[34] leads to a rapid decrease in the molten polymer viscosity and its molecular weight due to macromolecular chain cleavage. In comparison, PHBV copolymers are more suitable for the melt process because an increase in HV content results in lower melting and glass transition temperatures[35] and thus a wider processing window.

To improve such a drawback, or to give PHAs new properties, a great number of multiphase materials have been developed, mainly by mixing PHB or PHBV with other products such as plasticizers, fillers or other polymers.

6.2.2   PHA Plasticization

Many routes were investigated to ease PHB transformation[36] including plasticization. Many authors have noticed that PHA properties can evolve when plasticization occurs, *e.g.* with citrate ester (triacetin).[37–39]

Wang *et al.* tested different plasticizers—dioctyl phthalate, dioctyl sebacate, and acetyl tributyl citrate (ATBC)—with PHB. From the DSC measurements, they concluded that only the addition of ATBC leads to an obvious decline in T_g and improves other thermal characteristics. However, ATBC does little to improve the mechanical properties.[40]

The effects of biodegradable plasticizers on the thermal and mechanical properties of PHBV were studied by Choi and Park[41] using thermal and mechanical analyses. Soybean oil (SO), epoxidized soybean oil (ESO), dibutyl phthalate (DBP) and acetyl tributyl citrate (ATBC) were tested as plasticizing additives. PHBV/plasticizer blends were prepared by evaporating solvent from solutions. DPB and ATBC were more effective than soybean oils (SO and ESO) in lowering the glass transition temperatures as well as in increasing the elongation at break and the impact strength of the films. From the thermal and mechanical properties of the plasticized PHBV, it could be concluded that ATBC or DBP are better plasticizers for PHBV than SO and ESO.

From these different studies, it seems that acetyl tributyl citrate (ATBC) is the most efficient plasticizer for the main PHAs, *i.e.* PHB and PHBV.

6.2.3   PHA Nucleating Agents

Crystallization of the PHAs is often slower than desired and has a negative impact on the final properties of the material. It is thus important to ensure homogeneity and a suitable rate of crystallization after the processing of PHA for acceptable dimension stability of the article.

In this regard, certain compounds, called nucleating agents, are added to the PHA usually in an amount less than 1% (w/w) to perform two important roles: (1) speeding up the crystallization process during the cooling of the polymer matrix, and (2) leading to a greater number of smaller, homogeneously sized spherulites in the matrix.

The addition of nucleating agents leads to an increase in the crystallization temperature, crystallization rate, and crystallization enthalpy of the resulting PHA. The spherulites formed in the presence of nucleating agents are smaller than those formed by slow crystallization of the PHA polymer but since PHA has a low density of crystallization compared to polyethylene and polypropylene, the presence of nucleating agents also results in improved density and homogeneity of the crystalline domains in the polymer matrix. A variety of nucleating agents have been studied for PHB and PHBV polymers, such as boron nitride, talc, terbium oxide, and lanthanum oxide.[42] Other nucleating agents, such as calcium carbonate, hydroxyapatite, chalk, and micronized mica, have also been studied for PHB. The efficiency of a nucleating agent is found to vary with the PHA, for example, while boron nitride may be a good nucleating agent for PHB, the cyclohexyl phosphonic acid and zinc stearate nucleating agents are found to be better for PHBV, especially when the 3HV content is of a significant amount.[17] It was also reported that cyanuric acid was a better nucleating agent for PHBV and PHBH, reducing the crystallization half times by almost half and increasing the spherulite number along with a decrease in their size.[43] The physical interactions between the nucleating agent and the polymer affect the rate and the efficiency of the nucleating agent. Saccharin and phthalimide were found to be poorer nucleating agents for PHB and PHBV than boron nitride despite being soluble in the polymer melt. These two compounds were separated from the polymer matrix upon spherulite formation, occupying the regions in between the spherulites.[44] Some of the other interesting nucleating agents for PHA studied in the literature for biomedical applications are cyclodextrin,[45,46] lignin,[47] and orotic acid.[48]

The length of the chains in the polymer backbone and the presence of side chains affect the crystallization process of the PHAs; there is not a particular nucleating agent that is effective for all PHAs. However, a range of nucleating agents have been reported in the literature and based on the final application of the PHA, it may be advisable to choose one specific nucleating agent for a given PHA.

6.2.4 PHA Blends

Polymer blending offers the interesting possibility of preparing, at low cost, materials with adequate mechanical properties by combining the properties of two or more polymers. Several blending studies have been performed on PHB or PHBV copolymers with a range of compounds that can help vary the crystallinity and degradation of the final blends.[49,50] The possibility of blending other polymers with PHA offers the possibility of not only overcoming the drawbacks of a small processing window and low impact resistance of PHAs, but can also modify the crystallization tendency and biodegradation rates. Blending with hydrophilic polymers is known to increase the uptake of water and aid in faster biodegradation. The mechanical properties, morphology, biodegradability, and thermal and crystallization behaviour have been investigated on different PHBs blended with other polymers such as poly(vinyl alcohol), polysaccharides,[51,52] poly(ε-caprolactone),[53,54] poly(vinyl phenol), poly(vinyl acetate),[55] poly (lactic acid),[56] xylogen,[57] and even dendritic polyester oligomers.[58] PHA blends with other biodegradable polymers have been studied for specific applications also, for example, PHA-PCL blends were studied for cell adhesion and cell growth, rubber-PHA blends were studied for *e.g.* varying the thermal stability of the blends.

The possibility of H-bonding or formation of donor–acceptor interactions between PHA and the other blend constituents helps improve blend miscibility and reduces the tendency of phase separation.[50] To increase the compatibility and reduce the tendency of phase separation of the blends, reactive blending between the components has been developed. For instance, the components have been blended with peroxides. The latter decompose to form free radicals that cross-link the polymers[54] in the blend.

The miscibility and compatibility of these polyesters with functional polymers are well documented.[50,59]

6.2.5 PHA-based Multilayers

Multilayer coextrusion has been widely used, studied and understood in the domain of synthetic polymers for its effect on the viscosity of polymers, die geometry, layer distribution, encapsulation, and interfacial instabilities, influencing the quality and functionality of the multilayer products.[60–62] Despite the number and diversity of studies on multilayer flows and stability, only some articles report the use of biopolyester in coextrusion processes. Different stratified structures were processed by coextrusion and studied. But very few studies have been carried out with PHA. Most of these are based on the association between PHA *e.g.* PHBV, and plasticized starch.[60]

Applications of such PHA-based multilayers as commodities are primarily limited by PHA cost and have been until now by PHA availability, and thus attention is being focused on products with plastics constituting only a minor part, such as paper coatings like the plastic film moisture barrier in food and drink cartons and in sanitary napkins.

6.2.6 PHA Biocomposites

Biocomposites are obtained by the incorporation of macro-fillers (mainly ligno-cellulose fibres) in a biopolymer matrix. One of the main advantages of

PHA for biocomposite elaboration is its polar character. PHA shows better adhesion to ligno-cellulose fibres compared to conventional poly-olefins.[63] We can find a lot of papers based on PHA-biocomposites. The addition of cellulose fibres and different fillers has often been proposed as a solution to increase the mechanical performance and toughness of PHB and PHBV.[63–69]

In terms of crystallization and thermal behaviour, no significant effect of cellulose on PHB crystallinity was reported. A slight increase of T_g and a delay in the crystallization process were observed.[65] The presence of cellulose fibres also increases the rate of PHBV crystallization, due to a nucleating effect, while thermal parameters, such as crystallinity content, remained unchanged. Studies on the crystallization behaviour of PHB/kenaf fibre biocomposites showed that the nucleation by kenaf fibres affected the crystallization kinetics of the PHB matrix.[67] Differences in the effect of cellulose fibres on the crystallization process have been attributed to the lignin content at the surface/interface of the cellulose fibre.

The increase of HV content, the addition of compatibilizers and the rise of fibre content on PHA-based composites influenced the mechanical performance of the corresponding biocomposites.

For PHBV, the presence of HV led to a reduction in the stiffness but to increased elongation at break compared to PHB. In reinforced PHBV, a 50–150% enhancement in tensile strength, 30–50% in bending strength and 90% in impact strength have been reported.[66] The varying HV content in PHBV copolymers improved the toughness of the composites based on natural fibres and increased the ductility, but lowered the crystallization rate. It has been suggested, however, that the combination of coupling agents and HV units improved the storage modulus and led to a reduction in the tan δ,[63] due to an improvement in the interfacial bonding between PHA and the fibres and an increase in transcrystallinity near the fibre interfaces.

The addition of cellulose fibres led to some improvements in tensile strength and stiffness, but the composites remained brittle.[64] At low content, the incorporation of cellulose fibres lowered the stiffness, however, higher amounts of cellulose fibres greatly improved the mechanical properties of PHB.

For biocomposites based on cellulose fibres and PHB, the effects of fibre length and surface modification on the tensile and flexural properties have been investigated. The results on PHB reinforced with straw fibres have been published.[65] The fracture toughness values of composite materials containing 10–20 wt% straw fibres were higher than those of pure PHB, while biocomposites containing 30–50 wt% straw fibres presented about the same values as neat PHB.

With the addition of interface modifiers, the interfacial shear strength was also improved.[70] PHB containing wood flour and plasticizers presented a modest increase in tensile strength, while some improvement in terms of thermal stability was demonstrated.

6.2.7 PHA-based Nano-biocomposites

Nano-biocomposites are obtained by the incorporation of a nano-filler in a bio-matrix. In the literature,[71] nano-biocomposites based on poly-hydroxyalkanoates, mainly PHB and PHBV, were often prepared using different layered silicate clays (montmorillonite, sepiolite *etc*) and elaboration routes (casting, melt process).[72–75] Until now, a full exfoliation state has not been reported even if the beginning of a clay exfoliation has been shown in few studies.[72] Only intercalated or well intercalated structures and micro-composites were obtained using respectively organo-modified or un-modified layered silicates (clays). Despite the fact that fully exfoliated structures were not obtained, the mechanical and thermal properties, as well as the crystallization and biodegradation rates, were improved. Through many characterization techniques, regarding structural aspects as well as material properties, the role of nanoclays (type, content, and organization within the matrix) on PHAs' properties was highlighted and better understood.[76,77] The structure–property relationships for PHA/OMMT nano-biocomposites were established and are in good agreement with the conclusions drawn in previously reported studies on synthetic polymer-based nanocomposites. Further attention was paid to PHA degradation in nanocomposite systems since these polymers are very temperature sensitive.[74,75,78] A possible effect of the clay organomodifier and/or the mineral clay layer itself was pointed out, which can explain the limit of PHA improvements even with the addition of well dispersed nanoclays.

Considering these results, the poor thermal stability of PHAs and the effect of (organo-modified) clays on it represent major obstacles to the elaboration of technically competitive materials. Thus, scientists were interested in other PHA-based nanocomposites filled with *e.g.* layered double hydroxides (LDH),[79,80] cellulose whiskers[81–83] and hydroxyapatite (HA),[84] the latter being used in particular for biomedical and tissue engineering applications. LDH structures are similar to the layered silicate clays. Hsu *et al.* reached an exfoliated state using LDH organically modified by poly(ethylene glycol) phosphonates (PMLDH). The crystallization behaviour of these PHB/PMLDH were comparable to PHB/OMMT nanocomposites.[79,80]

In the case of PHA/cellulose whisker materials, studies were conducted using a latex of poly(3-hydroxyoctanoate) (PHO)[85] as a matrix and a colloidal suspension of hydrolyzed cellulose whiskers as natural and biodegradable fillers. Due to the geometry and aspect ratio of the cellulose whiskers, the formation of a rigid filler network, called the percolation phenomenon, was observed, leading to higher mechanical PHO properties.[81–83]

Eventually, regarding the nano-hydroxyapatite filler, like the polymer/clay nanocomposites, the good dispersion of inorganic fillers in the PHBV inevitably benefits the improvement of the mechanical properties of the materials.[84] Furthermore, the study also pointed out the enhanced material bioactivity since this specific property is expected for the repair and replacement of bone.

The use of graphene in a 2, 4, 6% composition in the PHBV led to an increase in the modulus at break of the PHBV composites. The improvement in the modulus with an increase in the graphene content was explained to be the result of the lamellar structures of graphene, which resulted in better polymer–filler interactions and therefore better stress transfer. The PHA–graphene composites, like the clay-nanocomposites of PHAs, were also reported to result in a certain level of thermal stability of the PHA. While in the case of clay nano-composites, the reduced thermal degradation is a result of reduced permeability of volatile degradation products, in the case of graphene reinforced composites, it was suggested to be the result of donor–acceptor complexes between graphene and the by-products of nucleophilic chain scission.[86]

"Green nanocomposites" based on PHAs appear as the next generation of environmentally-friendly materials and broaden the range of PHAs' applications by enhancing the polymer properties (ductility, melt viscosity, thermal stability). Thus, more appropriate new macromolecular architectures and nanoparticle-based systems should allow, in the near future, the limits of these materials (high crystallinity, brittleness, poor thermal stability) to be overcome.

6.3 Production and Some Applications

6.3.1 Industrial Production of PHA-based Materials

PHA production is shared between a great number of companies (see Table 6.2). Worldwide, several dozens of companies are known to be

Table 6.2 Main PHA producers.

Company	Country	Trade name	PHA	Pilot/industrial scale
Biomatera	Canada	Biomatera	PHBV	Pilot
Biomer	Germany	Biomer	PHB and copolymers	Pilot
Bio-On	Italy	Minerv PHA	PHB, PHBV	Pilot (?)
Kaneka	Japan	Kaneka	PHBHx	Pilot/ind. (?)
Meredian	US	?	Copolymers	Pilot, ind. (?)
Metabolix	US		Copolymers	Pilot, ind. (?)
PHB Industrial/ Copersucar	Brazil	Biocycle	PHB, PHBV	Pilot, ind. (?)
PolyFerm Canada	Canada	VersaMer PHA	PHBV and copolymers	Pilot
Tianan	China	Enmat	PHBV	Ind.
Tianjin & DSM	China	GreenBio	Copolymers based on 3HB and 4HB	Pilot (?)
Tianzhu	China	Tianzhu	PHBHx	Pilot

engaged in PHA production and applications.[87] Compared to PLA, the world production of PHA is low. But, it is difficult to have a precise idea because there is a wide gap between the (press) announcements and the true production. Most of the time, only the production capacities are given. In 2011, Nova-Institute (Germany) (http://www.nova-institut.de) published a market survey. According to this, the capacity of production of PHA was 51 150 tons per year in 2010. The corresponding previsions give 285 000 tons in 2016 and 405 100 tons in 2020.

The story of PHAs' industrial production is very complex and long, and begins in the 1950s. In the 1970s, Zeneca (formerly ICI) produced several metric tons of PHA copolymers under the trade name Biopol®. In the 1990s, Zeneca UK produced P(3HB-*co*-3HV) in a pilot plant by bacterial fermentation using a mixture of glucose and propionic acid. In 1996, Zeneca sold its Biopol® business to Monsanto, who continued investigations started by Zeneca into the production of PHA in genetically-modified crops. Monsanto commercially produced Biopol® P(3HB-*co*-3HV) with HV contents reaching 20% by means of fermentation. The production was ceased at the end of 1999. Metabolix bought Biopol® assets in 2001. In 2007, Metabolix and Archer Daniels Midland (ADM) formed a joint venture, Telles, to produce PHAs under the trade name Mirel. In 2012, this association was stopped. Beginning over 10 years ago, Metabolix continues the development of PHA. Metabolix has developed the production of PHA in genetically-modified crops. For instance, the company announced in 2009 that it has completed a field trial of tobacco, genetically engineered to express polyhydroxyalkanoate (PHA) bio-based polymers. This company has also announced that in greenhouse trials, switchgrass plants engineered using multi-gene expression technology can produce significant amounts of PHA in leaf tissues.

Different small companies currently produce bacterial PHA, *e.g.* PHB Industrial (Brazil) produces PHB and PHBV (HV = 12%) 45% crystalline, from sugar cane molasses.[88] Biocycle® production is planned to be increased to some thousand tons a year over the next few years. In 2004, Procter & Gamble (US) and Kaneka Corporation (Japan) announced a joint development agreement for the completion of R&D leading to the commercialization of Nodax, a large range of polyhydroxybutyrate-*co*-hydroxyalkanoates (PHBHx, PHBO, PHBOd).[89] Although the industrial large-scale production was planned with a target price of around 2€ per kg, the Nodax development was stopped in 2006.[90] In 2007, Meredian Inc. purchased P&G's PHA technology. Meredian plans to produce over 280 ktons in 2020. Tianan, a Chinese company, also announced that they plan to increase the capacity to some thousands a year in the future. The Dutch chemical company DSM announced its investment in a PHA plant together with a Chinese bio-based plastics company—Tianjin GreenBio-Science Co. The company will start up the production of PHA with an annual capacity of 10 000 tons. The Japanese company Kaneka plans to produce 10 000 metric tons per year in 2020.

6.3.2 The Case of PHA-based Materials for Biomedical Applications

The production of PHA is intended to replace synthetic non-degradable polymers for a wide range of applications: packaging, agriculture, leisure,[90] fast food, hygiene, as well as medicine and biomedical,[91,92] since PHA is biocompatible.

Biomedical applications require certain key properties such as purity, biocompatibility, cell growth and proliferation, and biodegradability to non-toxic products. An important consideration for the use of PHA for biomedical applications is the purity of the polymer as the usual processes of extraction and purification afford a PHA that still contains bacterial endotoxins (lipopolysaccharides/lipooligosaccharides), surfactant, and residual proteins, which do not qualify it as a pharmaceutical grade polymer. In this regard, purification of PHA is critical. The use of oxidizing agents using hydrogen peroxide, sodium hypochlorite, and benzoyl peroxide[92] has been proposed as a way to help remove the endotoxins, which are potential pyrogens, from PHAs. Treatment with supercritical fluids, *e.g.* supercritical carbon dioxide, helps in the removal of organic contaminants from the polymer. The combination of these methods affords PHAs of high purity suitable for pharmaceutical applications.[91]

Apart from being biocompatible, another important advantage of PHA over other polymers with potential in biomedical applications is the possibility of tailoring the properties from the wide range of building blocks, tuning the surface and mechanical properties, and the possibility of altering the degradation rate *in vivo*.[91,93] P(3HB-co-4HB) is particularly attractive for tissue engineering and implantation as it can be easily degraded by lipase in addition to PHA depolymerase. Thus, when used as a drug delivery agent, the susceptibility of the 4HB units to the lipase (hydrolytic enzyme) would make *in vivo* degradation possible even in the absence of PHA depolymerase. Other comonomers such as 5HV and 6HHx in PHA have also shown lipase-mediated degradation of the PHA.[15]

Different tissue engineering applications require specific degradation rates of the polymer. For example, implant-based applications, adhesion barriers and wound dressings require degradation of the polymer in a few weeks or months while other applications such as cardiovascular devices, stents and nerve repair devices would require very slow degradation lasting a few or several years.[94,95] In one study, the authors used UV radiation to facilitate the controlled degradation of PHBHx films meant for scaffolds and implants without significantly affecting the mechanical properties of the film. It was reported that the physical state in which the PHBHx was irradiated was important to the final properties of the film. When irradiation was carried out in powder form and then used for film preparation, the PHBHx showed mechanical properties similar to the films made from unirradiated PHBHx. It was also observed that UV treatment of PHBHx improved the hydrophilic character of the film surface allowing for improved cell adhesion and growth; whereas irradiation for the same duration and at the same intensity carried over the PHBHx films results in significant loss of the mechanical properties. Thus, depending on the end use of the PHAs, they can be modified using various methods with a certain degree of control over the final mechanical and surface properties.[93] An interesting feature of this study was also the decrease in the polydispersity index (molecular weight distribution) along with a decrease in the average molecular mass of the PHBHx polymer upon irradiation as seen from the SEC chromatograms. Thus, the length of the UV exposure can also be a parameter to control the final molecular weight distribution of the PHBHx polymer. PHAs also exhibit optical activity, antioxidant properties, and piezoelectric properties. As mentioned above, being a chiral polymer with an R configuration gives PHA very crystalline molecules. The crystalline structures of PHAs are found to have piezoelectric properties. The uniaxial lamellar crystals of PHB and PHBV have been reported to be deformed by the application of force that results in a change in the direction of their average dipole moment, a property called piezoelectric behaviour. The resulting dielectric constant and the piezoelectric strain constant (polarization/stress) were calculated at around a T_g of 15 °C.[96] The piezoelectric property is known to occur in various biopolymers (wood, bone, polysaccharides, proteins, deoxyribonucleates) but is not often seen in plastics and so makes PHAs an interesting class of polymers.[97] The piezoelectric properties of PHA have also contributed to its use in biomedical applications related to nerve repair. The use of PHB and PHBV have been studied for the preparation of guided nerve channels,[98] resorbable wrap-around implants to aid in nerve repair,[99] and conduits for nerve repair.[100] Other applications that may be partly related to the piezoelectric properties are its use as a bone filling augmentation material, ligament and tendon grafts[101,102] and so on.

Therefore, PHAs hold great promise in the field of biomedical sciences as a result of their many interesting physical characteristics.

References

1. R. Narayan, *Orbit Journal*, 2001, **1**, 1.
2. E. Pollet and L. Avérous, *Films and coatings from renewable resources – An applications perspective*, ed. D. Plackett, John Wiley & Sons, West Sussex, UK, 2011, ch. 4, pp. 65–86.
3. A. Steinbuchel, *Biopolymers, General Aspects and Special Applications*, Wiley-VCH, Weinheim, Germany, 2003.
4. M. Avella, E. Bonadies, E. Martuscelli and R. Rimedio, *Polym. Test.*, 2001, **20**, 517.
5. J. Fritz, U. Link and R. Braun, *Starch*, 2001, **53**, 105.
6. S. Karlsson and A.-C. Albertsson, *Polym. Eng. Sci.*, 1998, **38**, 1251.
7. D. L. Kaplan, J. M. Mayer, D. Ball, J. McMassie, A. L. Allen and P. Stenhouse, *Biodegradable Polymers and Packaging*, ed. C. Ching,

D. L. Kaplan and E. L. Thomas, Technomic Pub. Co, Lancaster, 1993, pp. 1–42.

8. A. Rouilly and L. Rigal, *J. Macromol. Sci., Polym. Rev.*, 2002, **C42**, 441.
9. K. Van de Velde and P. Kiekens, *Polym. Test.*, 2002, **21**, 433.
10. R. Chandra and R. Rustgi, *Prog. Polym. Sci.*, 1998, **23**, 1273.
11. M. C. Branciforti, M. C. S. Corrêa, E. Pollet, J. A. M. Agnelli, P. A. Nascente and L. Avérous, *Polym. Test.*, 2013, **32**, 1253.
12. J. M. B. Cavalheiro, E. Pollet, H. P. Diogo, T. Cesário, L. Avérous, M. C. M. D. Almeida and M. M. R. Fonseca, *Bioresour. Technol.*, 2013, **147**, 434.
13. M. G. E. Albuquerque, V. Martino, E. Pollet, L. Avérous and M. A. M. Reis, *J. Biotechnol.*, 2011, **151**, 66.
14. K. Sudesh, H. Abe and Y. Doi, *Prog. Polym. Sci.*, 2000, **25**, 1503.
15. J. A. Chuah, M. Yamada, S. Taguchi, K. Sudesh, Y. Doi and K. Numata, *Polym. Degrad. Stab.*, 2013, **98**, 331.
16. M. Akiyama, T. Tsuge and Y. Doi, *Polym. Degrad. Stab.*, 2003, **80**, 183.
17. J. Asrar and K. J. Gruys, *Biopolymers, Volume 4, Polyesters III - Applications and Commercial Products*, ed. Y. Doi and A. Steinbüchel, Wiley-VCH, Weinheim, Germany, 2002, pp. 57–67.
18. Y. Kann, *Plastics Technol.*, 2013, **59**, 31.
19. B. Laycock, P. Halley, S. Pratt, A. Werker and P. Lant, *Prog. Polym. Sci.*, 2013, **38**, 536.
20. K. Sudesh and H. Abe, *Practical Guide to Polyhydroxyalkanoates-Crystalline and Solid State Structures of polyhydroxyalkanoates*, ed. K. Sudesh and H. Abe, Smithers Rapra Technology, Shropshire, UK, 2010, pp. 25–37.
21. M. Yokouchi, Y. Chatani, H. Tadokoro, K. Teranishi and H. Tani, *Polymer*, 1973, **14**, 267.
22. J. Zhang, K. Kasuya, T. Hikima, M. Takata, A. Takemura and T. Iwata, *Polym. Degrad. Stab.*, 2011, **96**, 2130.
23. H. Sato, Y. Ando, H. Mitomo and Y. Ozaki, *Macromolecules*, 2011, **44**, 2829.
24. N. Grassie, E. J. Murray and P. A. Holmes, *Polym. Degrad. Stab.*, 1984, **6**, 47.
25. N. Grassie, E. J. Murray and P. A. Holmes, *Polym. Degrad. Stab.*, 1984, **6**, 95.
26. N. Grassie, E. J. Murray and P. A. Holmes, *Polym. Degrad. Stab.*, 1984, **6**, 127.
27. M. Kunioka and Y. Doi, *Macromolecules*, 1990, **23**, 1933.
28. Y. Aoyagi, K. Yamashita and K. Doi, *Polym. Degrad. Stab.*, 2002, **76**, 53.
29. S.-D. Li, J. D. He, P. H. Yu and M. K. Cheung, *J. Appl. Polym. Sci.*, 2003, **89**, 1530.
30. H. Abe, *Macromol. Biosci.*, 2006, **6**, 469.
31. F. Carrasco, D. Dionisi, A. Martinelli and M. Majone, *J. Appl. Polym. Sci.*, 2006, **100**, 2111.
32. D. H. Melik and L. A. Schechtman, *Polym. Eng. Sci.*, 1995, **35**, 1795.

33. R. Renstad, S. Karlsoon and A.-C. Albertsson, *Polym. Degrad. Stab.*, 1997, **57**, 331.
34. D. H. S. Ramkumar and M. Bhattacharya, *Polym. Eng. Sci.*, 1998, **38**, 1426.
35. W. Amass, A. Amass and B. Tighe, *Polym. Int.*, 1998, **47**, 89.
36. N. C. Billingham, T. J. Henman and P. A. Holmes, *Dev. Polym. Degrad*, 1987, **7**, 81.
37. M. A. Kotnis, G. S. O'Brien and J. L. Willett, *J. Environ. Polym. Degrad.*, 1995, **3**, 97.
38. R. L. Shogren, *J. Environ. Polym. Degrad.*, 1995, **3**, 75.
39. R. C. Baltieri, L. H. I. Mei and J. Bartoli, *Macromol. Symp.*, 2003, **197**, 33.
40. L. Wang, W. Zhu, X. Wang, X. Chen, G.-Q. Chen and K. Xu, *J. Appl. Polym. Sci.*, 2008, **107**, 166.
41. J. S. Choi and W. H. Park, *Polym. Test.*, 2004, **23**, 455.
42. J. W. Liu, H. L. Yang, Z. Wang, L. S. Dong and J. J. Liu, *J. Appl. Polym. Sci.*, 2002, **86**, 2145.
43. P. Pan, G. Shan, Y. Bao and Z. Weng, *J. Appl. Polym. Sci.*, 2012, **129**, 1374.
44. R. E. Withey and J. N. Hay, *Polymer*, 1999, **40**, 5147.
45. Y. He and Y. Inoue, *Biomacromolecules*, 2003, **4**, 1865.
46. Y. He and Y. Inoue, *J. Polym. Sci. Part B: Polym. Phys.*, 2004, **42**, 3461.
47. W. Kai, Y. He, N. Asakawa and Y. Inoue, *J. Appl. Polym. Sci.*, 2004, **94**, 2466.
48. N. Jacquel, K. Tajima, N. Nakamura, T. Miyagawa, P. Pan and Y. Inoue, *J. Appl. Polym. Sci.*, 2009, **114**, 1287.
49. J.-C. Huang, A. S. Shetty and M.-S. Wang, *Adv. Polym. Technol.*, 1990, **10**, 23.
50. H. Verhoogt, B. A. Ramsay and B. D. Favis, *Polymer*, 1994, **35**, 5155.
51. L. Finelli, M. Scandola and P. Sadocco, *Macromol. Chem. Phys.*, 1998, **199**, 695.
52. T. G. Cárdenas, J. L. Sanzana and L. H. I. Mei, *Bol. Soc. Chil. Quim.*, 2002, **47**, 529.
53. K. Sriroth and K. Sangseethong, *Acta Hortic*, 2006, **703**, 145.
54. F. Gassner and A. J. Owen, *Polymer*, 1994, **35**, 2233.
55. M. Grimaldi, B. Immirzi, M. Malinconico, E. Martuscelli, G. Orsello, A. Rizzo and G. M. Volpe, *J. Mater. Sci.*, 1996, **31**, 6155.
56. Y. An, L. Li, L. Dong, Z. Mo and Z. Feng, *J. Polym. Sci. Part B: Polym. Phys.*, 1999, **37**, 443.
57. L. M. W. K. Gunaratne and R. A. Shanks, *Polym. Eng. Sci.*, 2008, **48**, 1683.
58. K. Weihua, Y. He, N. Asakawa and Y. Inoue, *J. Appl. Polym. Sci.*, 2004, **94**, 2466.
59. S. Xu, R. Luo, L. Wu, K. Xu and G.-Q. Chen, *J. Appl. Polym. Sci.*, 2006, **102**, 3782.
60. R. Sharma and A. R. Ray, *J. Macromol. Sci., Polym. Rev.*, 1995, **35**, 327.

61. O. Martin, E. Schwach, L. Averous and Y. Couturier, *Starch/Staerke*, 2001, **53**, 372.
62. P. D. Anderson, J. Dooley and H. E. H. Meijer, *Appl. Rheol.*, 2006, **16**, 198.
63. J. Dooley, K. S. Hyun and K. Hughes, *Polym. Eng. Sci.*, 1998, **38**, 1060.
64. R. A. Shanks, A. Hodzic and S. Wong, *J. Appl. Polym. Sci.*, 2004, **91**, 2114.
65. P. Gatenholm, J. Kubat and A. Mathiasson, *J. Appl. Polym. Sci.*, 1992, **45**, 1667.
66. M. Avella, E. Martuscelli and B. Pascucci, *J. Appl. Polym. Sci.*, 1993, **49**, 2091.
67. A. K. Mohanty, M. Misra and G. Hinrichsen, *Macromol. Mater. Eng.*, 2000, **276–277**, 1–24.
68. M. Avella, G. Bogoeva-Gaceva, A. Buzoarvska, M. E. Errico, G. Gentile and A. Grozdanov, *J. Appl. Polym. Sci.*, 2007, **104**, 3192.
69. V. P. Cyras, M. S. Commisso, N. A. Mauri and A. Vazquez, *J. Appl. Polym. Sci.*, 2007, **106**, 749.
70. N. M. Barkoula, S. K. Garkhail and T. Peijis, *Ind. Crops Prod.*, 2010, **31**, 34.
71. S. Wong, R. A. Shanks and A. Hodzic, *Compos. Sci. Technol.*, 2007, **67**, 2478.
72. P. Bordes, E. Pollet and L. Avérous, *Nano- and Biocomposites*, ed. F. Hussain, A. Kin-tak Lau and K. Lafdi, CRC Press, Taylor & Francis Group, Boca Raton, FL, US, 2009, ch. 8, pp. 193–225.
73. P. Bordes, E. Pollet, S. Bourbigot and L. Averous, *Macromol. Chem. Phys.*, 2008, **209**, 1474.
74. P. Bordes, E. Pollet and L. Averous, *Prog. Polym. Sci. (Oxford)*, 2009, **34**, 125.
75. L. Cabedo, D. Plackett, E. Gimenez and J. M. Lagaron, *J. Appl. Polym. Sci.*, 2009, **112**, 3669.
76. L. Avérous and E. Pollet, *Environmental Silicate Nano-biocomposites*, Springer-Verlag, London, 2012, pp. 1–447.
77. M. C. S. Corrêa, M. C. Branciforti, E. Pollet, J. A. M. Agnelli, P. A. P. Nascente and L. Avérous, *J. Polym. Environ.*, 2012, **20**, 283.
78. P. Bordes, E. Hablot, E. Pollet and L. Averous, *Polym. Degrad. Stab.*, 2009, **94**, 789.
79. E. Hablot, P. Bordes, E. Pollet and L. Averous, *Polym. Degrad. Stab.*, 2008, **93**, 413.
80. T. M. Wu, S.-F. Hsu and C.-S. Liao, *J. Polym. Sci., Part B: Polym. Phys.*, 2006, **44**, 3337.
81. S.-F. Hsu, T. M. Wu and C.-S. Liao, *J. Polym. Sci., Part B: Polym. Phys*, 2007, **45**, 995.
82. D. Dubief, E. Samain and A. Dufresne, *Macromolecules*, 1999, **32**, 5765.
83. A. Dufresne, M. B. Kellerhals and B. Witholt, *Macromolecules*, 1999, **32**, 7396.
84. A. Dufresne, *Compos. Interfaces*, 2000, **7**, 53.

85. D. Z. Chen, C. Y. Tang, K. C. Chan, C. P. Tsui, H. F. Yu, M. C. P. Leung and P. S. Uskokovic, *Compos. Sci. Technol.*, 2007, **67**, 1617.
86. A. Dufresne and E. Samain, *Macromolecules*, 1998, **31**, 6426.
87. V. Sridhar, I. Lee, H. H. Chun and H. Park, *Polym. Lett.*, 2013, 7, 320.
88. U. Breuer, *Plastics from Bacteria: Natural Functions and Applications Series: Microbiology Monographs*, ed. Chen G. Q. G., Springer-Verlag, Heidelberg, Berlin, 2010, 14, 450.
89. A. El-Hadi, R. Schnabel, E. Straube, G. Muller and S. Henning, *Polym. Test.*, 2002, **21**, 665.
90. I. Noda, P. R. Green, M. M. Satkowski and L. A. Schechtman, *Biomacromolecules*, 2005, **6**, 580.
91. S. Philip, T. Keshavarz and I. Roy, *J. Chem. Technol. Biotechnol.*, 2007, **82**, 233.
92. S. F. Williams, D. P. Martin, D. M. Horowitz and O. P. Peoples, *Int. J. Biol. Macromol.*, 1999, **25**, 111.
93. M. Zinn, B. Witholt and T. Egli, *Adv. Drug Delivery Rev.*, 2001, **53**, 5–21.
94. Y. Y. Shangguan, Y. W. Wang, Q. Wu and G. Q. Chen, *Biomaterials*, 2006, **27**, 2349.
95. R. Rai, T. Keshavarz, J. A. Roether, A. R. Boccaccini and I. Roy, *Mater. Sci. Eng. R*, 2011, **72**, 29.
96. S. P. Valappil, S. K. Misra, A. R. Boccaccini and I. Roy, *Expert Rev. Med. Devices*, 2006, **3**, 853.
97. E. Fukada and Y. Ando, *Int. J. Biol. Macromol.*, 1986, **8**, 361.
98. E. Fukada, *Biorheology*, 1995, **32**, 593.
99. M. Borkenhagen, R. C. Stoll, P. Neuenscshwander, U. W. Suter and P. Aebischer, *Biomaterials*, 1998, **23**, 2155.
100. A. Hazari, G. Johansson-Rudén, K. Junemo-Bostrom, C. Ljungberg, G. Terenghi, C. Green and M. Wiberg, *Journal of Hand Surgery British and European Volume*, 1999, **24**, 291.
101. Y. Z. Bian, Y. Wang, S. Guli, G. Q. Chen and Q. Wu, *Biomaterials*, 2009, **30**, 217.
102. G. Q. Chen and Q. Wu, *Biomaterials*, 2005, **26**, 6565.

CHAPTER 7

Modification of Polyhydroxyalkanoates (PHAs)

A. M. GUMEL, M. H. ARIS AND M. S. M. ANNUAR*

Institute of Biological Sciences, Faculty of Science, University of Malaya, 50603 Kuala Lumpur, Malaysia
*Email: suffian_annuar@um.edu.my

7.1 Introduction

Polyhydroxyalkanoates (PHAs) are biodegradable and biocompatible polyesters with versatile structural compositions. Bacterial PHAs are produced using a combination of renewable feedstock and biological methods mostly *via* a fermentation process. Native and recombinant microorganisms have been generally used to produce different types of PHAs, such as homopolymers[1,2] and copolymers of diverse morphology.[3–5] Alternative production schemes of PHAs *in vitro* based on cell-free enzymatic catalysis are gaining momentum and may become the preferred route to some specialty products.[6,7]

In addition to their biodegradability, compatibility, and compostability, PHAs were reported to possess gas-barrier properties almost similar to those of polyvinyl chloride and polyethylene terephthalate.[8] These combinations of excellent physico–chemical properties coupled with the current concerns over environmental pollution and waste degradation drive their increasing commercial exploitation in different niche applications spanning from biomedical, packaging, automotive, infrastructure, aerospace to military applications.[7,9]

RSC Green Chemistry No. 30
Polyhydroxyalkanoate (PHA) Based Blends, Composites and Nanocomposites
Edited by Ipsita Roy and Visakh P M
© The Royal Society of Chemistry 2015
Published by the Royal Society of Chemistry, www.rsc.org

Unfortunately, despite their high potential for commercial applications, most of the PHAs produced, especially those belonging to crystalline short chain length PHAs or medium chain length PHAs with higher monomeric compositions of 3-hydroxybutyric acid, were said to be highly hydrophobic, and exhibited brittleness, a low heat distortion temperature and poor gas-barrier properties. This resulted in slow degradability and resorbability within the extracellular matrixes as well as limited malleability and ductility during industrial processing.[10,11] As such, these kinds of neat biopolymers fail to meet the industrial demands especially in aggressive environments. Therefore, several approaches were devised to enhance the physico–chemical properties of the PHA in order to overcome these shortcomings by modifying the biopolymers through different processes.[9]

For example, metabolic engineering and culture condition manipulation were employed to produce modified PHAs with salient features. PHAs with methyl side-chains such as PH6N (poly(3-hydroxy-6-methyl-nonanoate)) was obtained from *Pseudomonas oleovorans* fed with methylated alkanoic acids or in a mixture with nonanoic acid as a carbon source.[12] PH6N crystallizes much faster than neat PHN (poly-3-hydroxynonanoate),[13] and the melting temperature was higher ($T_m = 65$ °C) than that of PHN ($T_m = 58$ °C).[14] Studies have shown that PHAs containing an epoxidized group were accumulated inside *P. oleovorans* when fed with octanoate and 10-undecenoic acid[15] and inside *P. cichorii* YN2 when fed with 1-heptene to 1-dodecene as a sole carbon source.[16] Alternatively, *P. oleovorans* accumulated sulfanyl PHA upon feeding with *m*-(*n*-thienylsulfanyl) alkanoic acid.[17] Furthermore, the bacteria was found to accumulate brominated PHA when fed with a mixture of ω-bromoalkanoic acids and nonanoic or octanoic acid.[18] Changing the carbon source to octane and 1-chlorooctane or nonanoic acid and fluorinated acid co-substrates resulted in the accumulation of chlorinated or fluorinated PHA in *P. oleovorans,* respectively.[19,20] Feeding propylthiooctanoic acid (PTO) or propylthiohexanoic acid to metabolically engineered *Ralstonia eutropha* led to the production of thiol functionalized PHA with enhanced chemical properties.[21] On the other hand, addition of low molecular weight diols such as ethylene glycol to the fermentation medium resulted in the bacterium producing a bifunctional telechelic hydroxyl-terminated PHA.[22,23]

The presence of these types of functional groups in specialized PHAs allows for further functional group modification. However, such biosynthetic approaches could only produce the specified polymer in minute quantities and most of the time, the whole-cell metabolic framework of the fermentation process itself limits the degree of freedom in designing the PHA with other functional groups of interest. As such, alternative routes for modification need to be applied. In order to extend their applications in aggressive environments where the neat polymers have failed, modification (functionalization) of the polymer *via* chemical,[24] physical[25] or enzymatic[26] processes have been employed. For example, sugars with pyranose structures, such as galactose and mannose, were reported to be ligands specific to the over-expressed asialoglycoprotein receptor (ASGPR) in hepatocellular carcinoma.[27] Hence, modifying the PHA *via* attachment with these types of sugars could help to enhance the cancer drug targeting potential of the modified material.[28] Similarly, chemically modified PHA with folic acid could efficiently serve as a cancer drug carrier by targeting folate receptor (FR), a glycosylphosphatidylinositol anchored protein, which appears to be up-regulated in more than 90% of non-mucinous ovarian carcinomas.[24]

Modification of PHA pendant groups was reported to enhance the degradation rate of the polymer.[29] For example, converting the pendant groups to carboxylic acids or basic amine groups enhances PHA degradation by pH alteration. On the other hand, pendant groups converted to hydrophilic groups (2-hydroxy acids *e.g.* glycolic and lactic acid) increase the uptake of hydrolytic agents such as water or a physiological buffer. Moreover, the pendant groups may also be converted to groups that would increase the polymer's porosity using active compounds such as inorganic salts and sugars that are removed by leaching. In addition, modifying PHAs by direct fluorination was reported to cause a marked change in their thermo–chemical properties thereby extending their niche applications.[30] Other reactive groups, such as amines, alcohols, amino acids and amino alcohols, and multifunctional monomers such as triols and tetraols, were also used as polymer pendant modifiers that can subsequently participate in polymer intra- or inter-molecular modifications.[29] As well as expanding their applications, modifications to PHAs also increase the availability of different types of biologically produced polymers. This chapter discusses the different methods of PHA modification.

7.2 Chemical Modification of PHAs

The structure of a PHA can be altered chemically to produce a modified polymer with predictable variation in molecular weight and functionality. For example, the hydrolytic rate of PHA to give an activated macromer that can further accept a reactive functional group is said to depend on several factors such as the chemical nature or reactivity of the ester linkages between the monomers.[29] Thus, the hydrolytic rate can be modified to become relatively fast by incorporating more hydrolysis-prone chemical groups (*e.g.* hydroxy-acids such as 2-hydroxyethoxy acetic acid, amide, anhydride, carbonate, carbamate *etc*) into the polymeric backbone thereby modulating the reactivity of the ester linkage.[23]

Unlike PHA modification *via* manipulation of the fermentation process, chemical reactions allow for the bulk production of a uniform product and incorporation of diverse functional groups to produce useful tailor-made polymers with desirable properties for niche applications. PHA modification by chemical processes can be achieved by different methods such as carboxylation, hydroxylation, epoxidation, chlorination, grafting reaction *etc*. (Figure 7.1).[14,31]

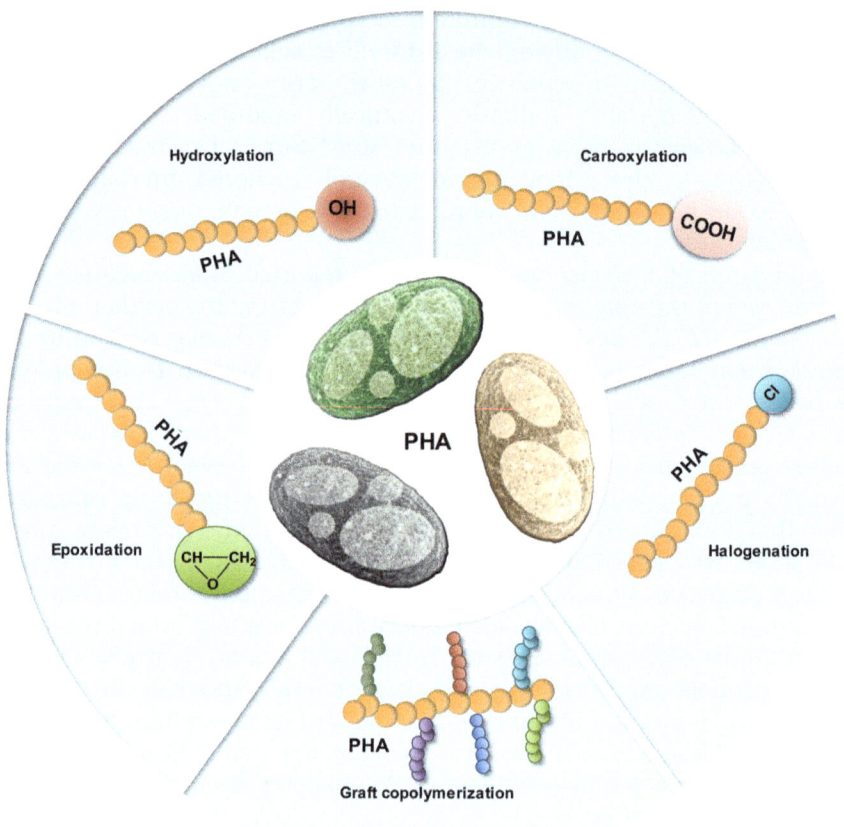

Figure 7.1 Typical methods for the chemical modification of PHAs to yield different types of functionalized polymers.

7.2.1 Carboxylation

PHA modification by carboxylation is the addition of carboxylic functional group to the polymeric macromer (Figure 7.2). Carboxylic groups incorporated into the polymer usually serve as functional binding sites for bioactive moieties such as probes for targeting proteins and hydrophilic components.[31] Studies have shown that the carboxylation of PHA enhances the hydrophilicity of the polymer through better water penetration, which is encouraged *via* ester group hydrolysis by water.[32–34]

A chemical oxidative process is normally employed to carboxylate PHA with an unsaturated group.[14,31,33] Previously, de Koning *et al.*[35] carboxylated the double bond of poly-3-hydroxyoctanoate-*co*-3-hydroxyundecenoate (PHOU) using KMnO$_4$ as an oxidation agent in the presence of NaHCO$_3$. This method was further modified by Kurth *et al.*[34] and Lee and Park[32] where crown ether was used as the phase transfer and dissociating agent for the

Figure 7.2 Oxidative carboxylation of unsaturated PHA using KMnO$_4$ as an oxidation agent.

KMnO$_4$ to carboxylate the unsaturated PHOU (Figure 7.2). The authors were able to quantitate the oxidation based on the disappearance of the ^{1}H NMR peak ($\delta = 4.9$ and 5.7 ppm) assigned to the double bond.[34] Although a complete oxidation of the double bonds was achieved, the researchers found that 25% carboxylation of PHOU was sufficient to enhance the hydrophilicity of the polymer.[25] Furthermore, they reported that the carboxylated polymer suffered a major loss of molecular weight compared to neat PHOU.[32,34] The decrease in the molecular weight was attributed either to macromolecular chain degradation during the process or as a result of differences in hydrodynamic radii between the carboxylated polymer and SEC polystyrene standards.[34] Lee and Park[32] reported a reaction time of 2 h as optimal to achieve maximum carboxylation of 50%, where a modified PHA with a degree of carboxylation between 40–50% was observed to be completely soluble in water/Na$_2$CO$_3$.[32]

In an attempt to address the issue of molecular weight degradation upon carboxylation, Stigers and Tew[36] developed a new synthetic process that uses osmium tetraoxide and oxone in dimethylformamide. The oxidation reaction took place at 60 °C for 8 h, and proceeded to completion with little polymer degradation.[36] The carboxylic group in the modified PHA proved advantageous in producing block and grafted copolymers. Condensation reactions between carboxylic acids and amine groups were exploited to graft modified PHA and linoleic acid onto chitosan.[37] Recently, Babinot *et al.*[38] used click ligation to esterify the pendant –COOH of carboxylated PHA with propargyl alcohol resulting in a clickable-alkyne group that was subsequently used to copolymerize poly(ethylene glycol) (PEG) macromer onto the modified PHA.

7.2.2 Hydroxylation

The properties of PHA and its copolymers have been reported to be modified by hydroxylation.[39–42] Normally, acid- or base-catalyzed reactions are used in the modification of PHA by hydroxylation in the presence of low molecular weight mono- or diol compounds (Figure 7.3). Hydroxy-terminated PHA is of importance in block copolymerization. Methanolysis of PHA resulted in PHA methyl esters bearing monohydroxy-terminated groups.

Timbart *et al.*[43] reported the production of monohydroxylated oligomers of poly-3-hydroxyoctanoate (PHO) and PHOU using both base- and

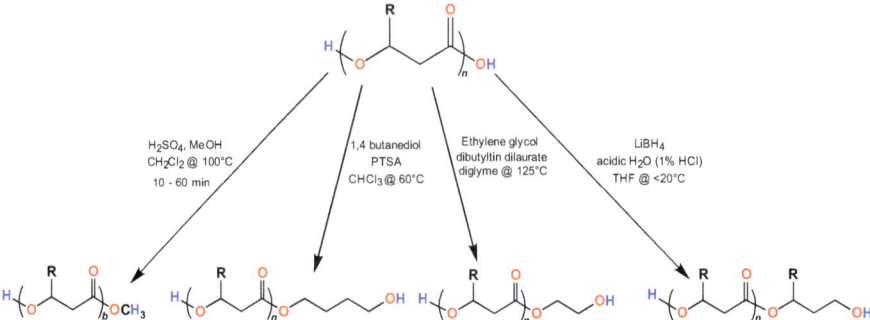

Figure 7.3   Different approach to PHA modification by hydroxylation; R = alkyl side chain.

acid-catalyzed hydrolyses. In basic hydrolysis, the researchers used alcoholic NaOH to catalyze the hydrolysis at pH 10–14 and the reaction was stopped upon addition of concentrated aqueous hydrochloric acid. On the other hand, the acid hydrolysis was performed using two different approaches *viz.* (i) a reaction catalyzed by *para*-toluenesulfonic acid monohydrate (PTSA) at 120 °C, which was stopped by cooling the mixture in an ice bath; (ii) monohydroxylation by acidic (sulfuric acid) methanolysis at 100 °C to yield the respective 3-hydroxymethyl esters bearing a methyl protected carboxylic acid group. It was also reported that the kinetics of PHO oligomer production depends on the reaction conditions. The ester bonds of PHO were shown to be stable within pH 10–12. Increasing the pH to 14 resulted in a higher hydrolysis rate and produced oligomers with unsaturated end groups due to a McLafferty rearrangement.[43] The researchers concluded that an acid-catalyzed reaction and methanolysis were more efficient than basic hydrolysis in producing PHO oligomers. They also showed that the observed decrease in polymer molecular weight was dependent upon the solvent type used. Polymeric chain cleavage was observed to occur more frequently in toluene than in dichloroethane, an observation that was attributed to the better solubility of the polymer in toluene. In another study, *p*-toluenesulfonic acid (PTSA) in the presence of methanol was used to catalyze the monohydroxylation of poly-3-hydroxybutyrate-*co*-4-hydroxybutyrate (P3HB4HB).[39] The modified PHA was acrylated and grafted onto poly(ethyleneimine) *via* Michael addition, which resulted in a material that was used to deliver siRNA (Figure 7.4).[39]

In the preparation of PHA tri-block copolymers, a bi-functional telechelic PHA as a macro-initiator is usually required. In this process, low molecular weight diols are used as the micro-initiator of the PHA-diol reaction. During the reaction, the hydroxyl groups of the diol were proposed to cleave the polymeric ester bonds in a random fashion, resulting in the dihydroxy terminated PHA.[31] Dibutyltin dilaurate was used as a chemical catalyst in the presence of ethylene glycol (micro-initiator) to produce an enantiomerically pure telechelic dihydroxy-terminated PHO and its

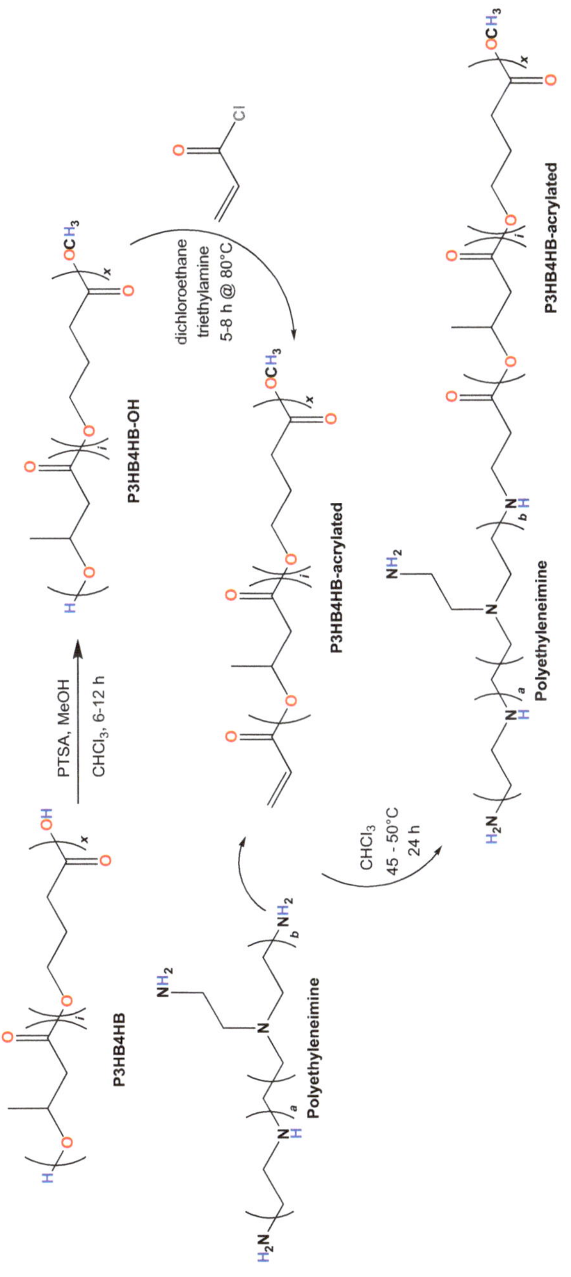

Figure 7.4 PHA modification by monohydroxylation and subsequent graft polymerization by Michael addition.

copolymers at 80–91% yield.[42] It was found that the modified telechelic PHO-diol showed lower glass transition (T_g) and melting (T_m) temperatures than neat PHO. However, this observation was found to contradict the reported increase in the thermal stability of di-hydroxy terminated P3HB synthesized by microbial fermentation.[23] Using 1,4-butanediol and PTSA, Chen *et al.*[44] produced modified P3HB4HB and poly-3-hydroxybutyrate-*co*-3-hydroxyhexanoate (PHBHHx) diols that were grafted with poly(ester-urethanes) *via* Michael addition. Recently, Kwiecień *et al.*[40] described a new method for P3HB4HB hydroxylation by selective partial degradation of the polymer ester bonds using lithium borohydride. The reaction was carried out by drop-wise addition of 2 M LiBH$_4$ at a temperature <20 °C after the polymer was completely solubilized in tetrahydrofuran (THF). After a series of purification steps of chloroform and water washing followed by evaporation of the organic phase, about 97% of purified modified PHA oligodiols were obtained. As revealed by NMR and electrospray ionization-mass spectrometry (ESI-MS) analyses, the applied reduction process of the PHA oligodiols was highly selective.[40] The researchers claimed the method to be versatile and applicable to the production of any PHA oligodiol. Block copolymers of PHO and poly(ester-urethanes) were synthesized by a reaction between the polymer hydroxyl group and the –NCO group of aliphatic *L*-lysine methyl ester diisocyanate (LDI) that was employed as a junction unit.[45] The aliphatic diisocyanate (LDI) was chosen over the aromatic diisocyanate due to the absence of the toxic, mutagenic, and carcinogenic aromatic amine degradation byproducts of the aromatic diisocyanate derived polyurethane. Similarly, Chen *et al.*[46] used facile melt polymerization to synthesize a block poly(ester-urethane) based on diols of P3HB4HB and poly-3-hydroxyhexanoate-*co*-3-hydroxyoctanoate (P3HHx3HO) using 1,6-hexamethylene diisocyanate (HDI) as the coupling agent. Lactate dehydrogenase (LDH) assay and platelet adhesion determination revealed that the synthesized material showed much higher platelet adhesion than the neat polymers, and even higher than that of polylactic acid (PLA) and poly(3-hydroxybutyrate) (PHB).[46] In fact, this type of poly(ester-urethane) was reported to possess the properties required for use in medical applications including enhanced wound healing activity.[46] A large group of PHA-based materials were produced from hydroxylated PHA, and have been reviewed extensively.[31]

7.2.3   Epoxidation

The high reactivity of epoxide groups under mild conditions was described as an important factor in medium-chain-length (mcl) PHA modification (Figure 7.5(a)). The epoxide group can participate in diverse reactions such as cross-linking (Figure 7.5(b)) to attach copolymers, bioactive moieties, or an ionizable group without unwanted polymer degradation.[31,47] Several studies have reported successful chemical modification of PHA by epoxidation.[48,49]

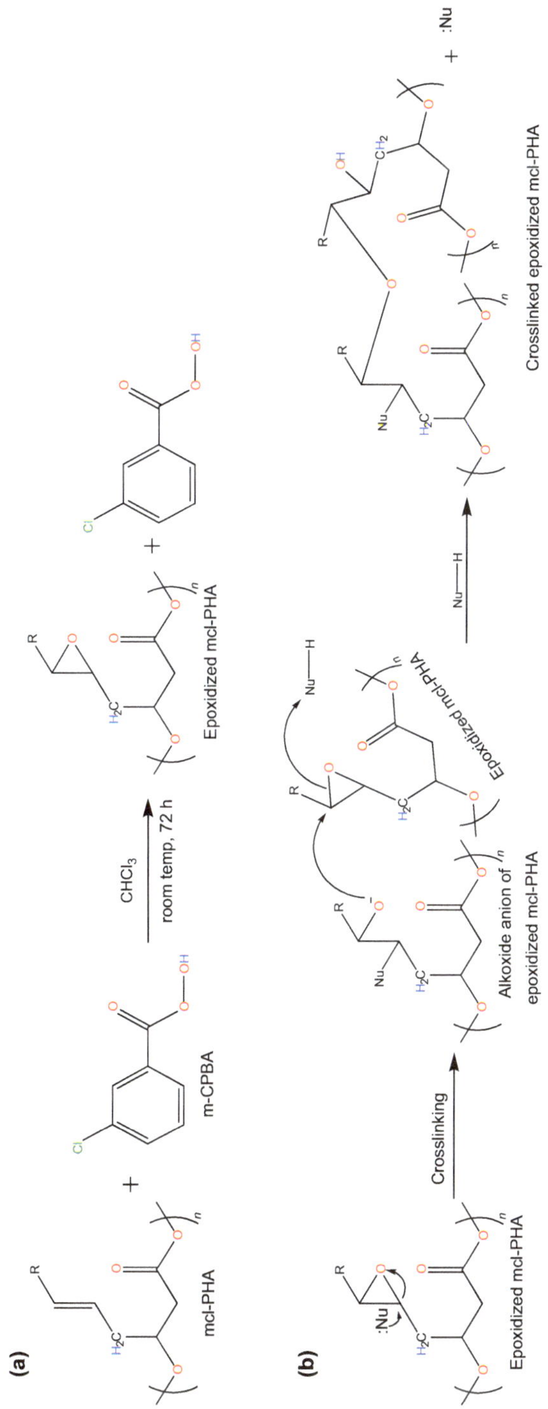

Figure 7.5 (a) Epoxidation of mcl-PHA using *m*-CPBA; (b) ether cross-linking of the epoxidized mcl-PHA.

A study by Bear *et al.*[15] reported the chemical epoxidation of mcl-PHA obtained from *P. oleovorans* and *Rhodospirillum rubrum* using *m*-chloroperoxybenzoic acid (*m*-CPBA). Proton NMR analysis revealed about 36.7% epoxidation that was calculated upon comparing the α,β-oxirane proton signals (2.75 and 2.9 ppm) with the signals of methylene protons (2.6–2.5 ppm) in the PHA backbone. Since Bear *et al.*'s report on the chemical epoxidation of PHA, there have been several reports on a similar epoxidation process. Park *et al.*[48] epoxidized PHOU with a controlled amount of olefinic bonds using *m*-CPBA.[48] Regardless of the number of polymeric olefinic groups, the researchers observed the process to follow second-order reaction kinetics with an observed initial reaction rate (v) of 1.1×10^{-3} L mol^{-1} s^{-1} at 20 °C. However, both the melting temperature (T_m) and melting enthalpy were observed to decrease with increasing conversion of the olefinic bonds to epoxy groups. Interestingly, the authors reported an increase in glass transition temperature (T_g) by about 0.25 °C for each 1 mol% of epoxide group, irrespective of the PHOU composition used. In similar studies, the researchers cross-linked the epoxidized PHOU with succinic anhydride in a reaction initiated by 2-ethyl-4-methylimidazole and carried out at 90 °C for a period of 0.5 to 4 h, which resulted in a highly elastic cross-linked PHA.[49] In their study, they found that by carrying out the reaction in mild acidic conditions, the reported polymer degradation was inhibited. The cross-linking kinetic parameters were evaluated using Kissinger[50] and Ozawa[51] models, and they calculated a cross-linking activation energy of 15.6–16.0 kcal mol^{-1} in all the reactions.[49] When evaluating the thermal stability of the epoxidized PHA, Park *et al.*[52] observed that the polymer thermal stability increases with increasing epoxy group. The observed increase in the thermal stability was attributed to intermolecular thermal cross-linking reactions between the pendant epoxy groups and the carboxylic acid groups generated from the polymer random chain scission by β-elimination (Figure 7.6). This interpretation was derived from the appearance of thermal exothermic peak "b" (375 °C) normally associated with a cross-linked reaction, followed by the endothermic melting temperature peak "a" (299 °C) in the differential scanning calorimetric (DSC) thermogram of the epoxidized polymer (Figure 7.7).

Mcl-PHA obtained from linseed oil was reported to possess a high number of olefinic side-chains, making the polymer consistently viscous and sticky at room temperature.[53] The polymer has limited potential applications, except as a bio-adhesive, but the range of applications can be expanded by improving the rigidity and stiffness of the polymer. Ashby *et al.*[53] used *m*-CPBA (as illustrated in Figure 7.5(a)) to convert about 37% of olefinic bonds in linseed oil derived mcl-PHA side-chains to epoxy groups in order to enhance the mcl-PHA cross-linking ability.

Comparing the ^{13}C NMR spectra of both the neat mcl-PHA (Figure 7.8(a)) and the epoxidized mcl-PHA (Figure 7.8b), the researchers confirmed the polymer epoxidation by the appearance of an epoxide chemical shift at 58 ppm as shown in Figure 7.8(b). They explained that steric hindrance

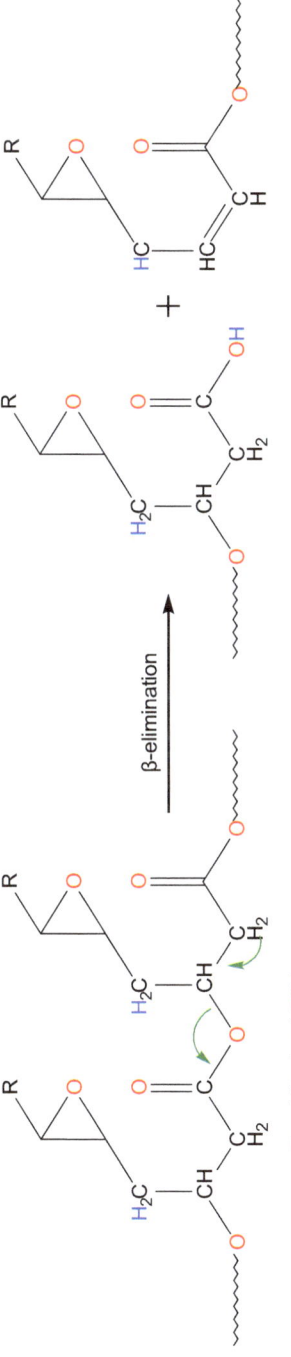

Figure 7.6 Generation of carboxylic acid from polymer random chain scission by β-elimination.

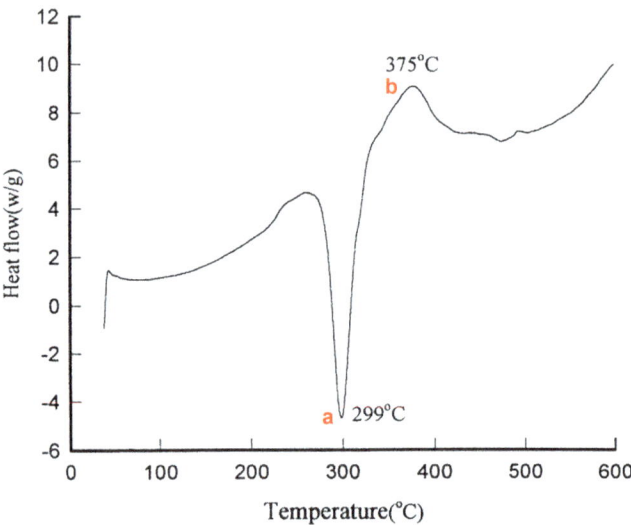

Figure 7.7   DSC thermogram for epoxidized PHO showing an endothermic melting
peak (a) and exothermic cross-linking peak (b) reproduced from Park
et al.[52] with permission from Elsevier.

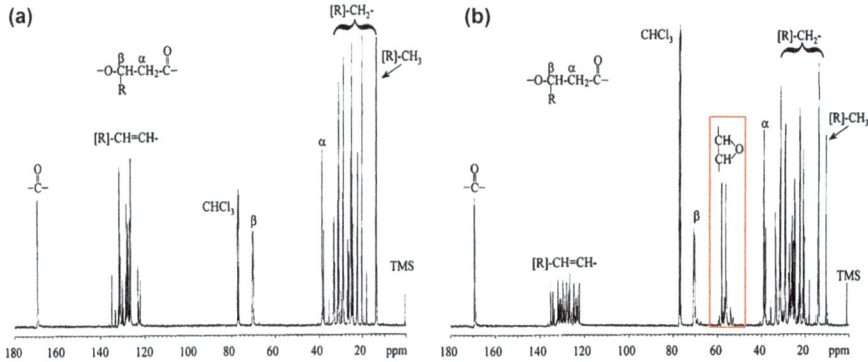

Figure 7.8 ^{13}C NMR spectra of linseed derived mcl-PHA (a) neat polymer (b)
epoxidized polymer, reproduced from Ashby *et al.*[53] with permission
from Elsevier.

caused by the proximity of the internal olefins to the side-chains to be the
reason behind the low yield of olefinic conversion. The intermolecular cross-
linking was suggested to have occurred *via* ether cross-links as depicted in
Figure 7.5(b). In this type of cross-linking, it was proposed that a nucleophile
(Nu) or radical species first opened the carboxy-oxirane ring leading to rearrangement of electrons and the formation of an alkoxide anion, which in
turn initiated a nucleophilic attack on another epoxidized PHA oxirane
carbon (Figure 7.5(b)).[53]

The study showed that the epoxidation enhanced the cross-linking of the mcl-PHA when exposed to air, resulting in an increased tensile strength and Young's modulus from 4.8 to 20.7 and 12.9 to 510.6 MPa, respectively. Consequently, this conferred short-term (25 days) stiffness to the modified PHA compared to a long period (50–75 days) for the corresponding neat PHA.[53]

In another study, Lee *et al.*[54] reported a successful cross-linking of epoxidized PHO using hexamethylene diamine (HMDA) as a cross-linker in a reaction carried out at 90 °C for 0.5 to 24 h (Figure 7.9). The degree of cross-linking was found to significantly influence the T_g and relative storage modulus of the modified copolymer. In addition, the effect of cross-linking on the thermal stability of the epoxidized PHA was studied by cross-linking the PHA with either succinic acid (in the presence of a catalyst) or HMDA (in the absence of a catalyst).[55] The study showed that the increment in the amount of basic catalyst or diamine cross-linker caused a decline in the thermal stability of cross-linked PHA.[55] The observation was attributed to ester cleavage of the polymer backbone catalyzed by the basic catalyst or cross-linker amine groups.

7.2.4 Halogenation

Halogenation of PHA is considered an excellent method in diversifying the polymer functions and applications. Halogen atoms such as chlorine, bromine and fluorine were added to the olefinic bonds of unsaturated PHA through an addition reaction,[56] and to the saturated PHA *via* substitution reactions.[57] Excess HCl was added to KMnO$_4$ in a drop wise fashion to generate chlorine gas (Figure 7.10). The gas was subsequently passed into a solution of sticky unsaturated PHA obtained from *P. oleovorans.* This resulted in a stiff and crystalline polymer at approximately 54 wt% chlorination.[56] Depending on the chlorine content, the chlorinated PHA (PHA-Cl) exhibited higher melting and glass transition temperatures ($T_m = 125$ °C, $T_g = 58$ °C) compared to neat PHA ($T_m = 55$ °C, $T_g = -50$ °C).[56] However, an observed hydrolysis of the polymer backbone was also reported due to the reduced molecular weight of the PHA-Cl with increasing chlorine content.[56]

In another study, Arkin and Hazer[57] modified the PHA-Cl into quaternary ammonium salts, thiosulfate moieties and phenyl derivatives. In addition, they cross-linked the modified PHA-Cl with benzene by electrophilic aromatic substitution using a Friedel–Crafts reaction. The random composition of PHA-Cl was calculated from its ^{1}H NMR spectrum by comparing the relative peak areas of the methine protons on the polymer backbone. Hence, increased chlorination of the methyl protons caused the peak of methine protons to be moved further downfield. In addition, the PHA-Cl mole fractions were calculated by comparing the peak areas of protons on chlorinated α-carbons and protons on β-carbons.[57] Samsuddin *et al.*[30] described a process for the direct fluorination of PHBHHx at elevated pressure in the

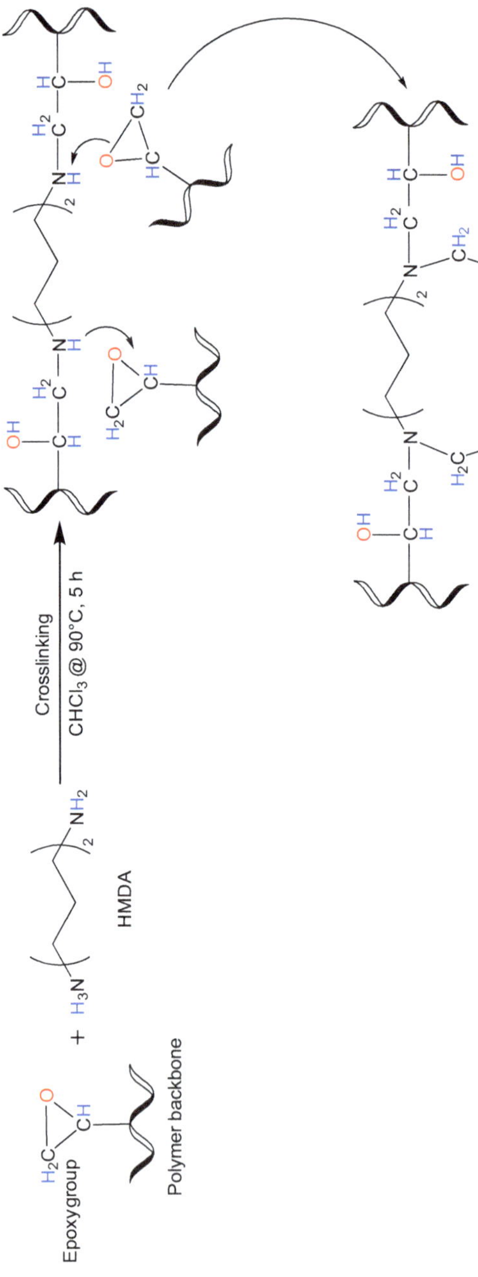

Figure 7.9 Cross-linking of epoxidized mcl-PHA using hexamethylene diamine (HMDA) as a cross-linker.

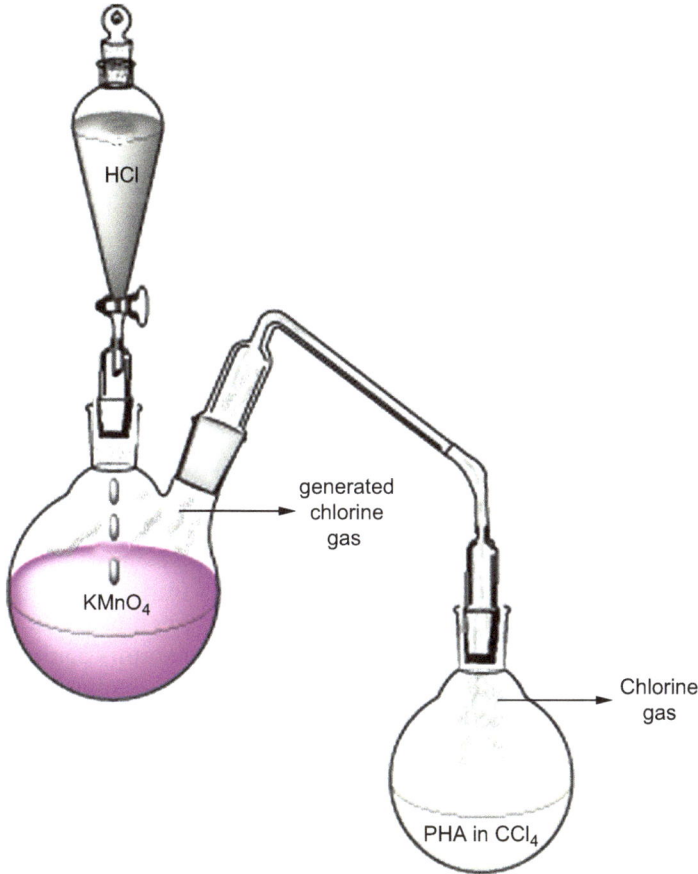

Figure 7.10 Reaction set-up for the chemical modification of PHA *via* chlorination.

presence of an elemental F_2/N_2 gas mixture. The fluorinated polymer was reported to show marked changes in its thermo–chemical characteristics.[30]

Mihara *et al.*[58] and Imamura *et al.*[59] filed an embodiment that detailed the procedures for PHA chemical modification by sulfanyl halogenation and the potential application of the modified PHA as a toner electrostatic charge controller in electrophotographic imaging. The inventors disclosed that the modified PHA possesses excellent charging stability, high charging ability, and enhanced dispersibility, when used as a toner charge control agent in an electrophotographic process.[58,59]

7.2.5 Graft and Cross-linking Polymerization

Another method to modify PHA is by graft copolymerization, which results in the formation of a modified segmented copolymer with improved properties such as increased wettability and thermo–mechanical strengths.

Grafting reactions can be induced by either chemical, radiation or plasma discharge methods.[60,61] Among the different approaches used in chemical graft copolymerization, the "grafting through" approach is by far the simplest process. The approach involves copolymerizing the PHA macromer with low molecular weight monomer by atom transfer radical polymerization to obtain a well defined copolymer with improved physico–chemical properties. Lao *et al.*[62] grafted 2-hydroxyethylmethacrylate (HEMA) onto poly(3-hydroxybutyrate-*co*-3-hydroxyvalerate) (PHBHV) using benzoyl peroxide (BPO) as a micro-initiator. The study revealed that grafting HEMA onto the PHA helped to improve the crystallinity and wettability of the polymer. Similarly, methyl methacrylate (MMA) was thermally grafted onto olefinic PHA derived from soybean oil using BPO as a chemical initiator.[63] A fractional precipitation method was used to isolate the graft copolymer from the reaction mixture of related homopolymers. The material was stored in hydroquinone to prevent intermolecular post polymerization by cross-linking. The modified polymer was shown to exhibit improved thermo–mechanical properties.[63] Higher antimicrobial activity and biocompatibility were imparted into PHO by grafting with vinylimidazole using the same thermal grafting approach in the presence of BPO initiator, resulting in a modified polymer (VI-*g*-PHO). The grafted polymer, when tested at 2% (w/v) suspension, showed a >90% reduction in viable cell counts against *Escherichia coli, Staphylococcus aureus* and *Candida albicans*.[64] Similarly, an ozone-treated PHBV membrane was grafted with acrylic acid that was subsequently esterified with chitosan. A collagen immobilized membrane was shown to improve L929 murine fibroblast cell proliferation. It also exhibited enhanced antimicrobial activity against *Staphylococus aureus, Escherichia coli* and *Pseudomonas aeruginosa*.[65] Application of PHA in neoglycoconjugate reactions helps to improve its hydrophilicity, degradability and bioresorbability. The conjugated PHAs were reported to be highly important in biological studies such as tissues and stem cell research.[66] Anti-Markovnikov addition to the double bond of unsaturated PHA was used to graft maltosyl units resulting in a hydrophilic grafted copolymer that is readily soluble in dimethylformamide and dimethyl sulfoxide but insoluble in dichloromethane and chloroform.[67] When acrylic acids were grafted onto ozone modified PHB and PHBV, followed by grafting of chitosan or chitooligosaccharide *via* esterification,[68] the modified PHA showed a strong antibiotic activity against *Escherichia coli, Pseudomonas aeruginosa, Staphylococcus aureus* and methicillin-resistant *S. aureus* (MRSA). In addition, acrylic acid grafting alone increases the biodegradability, whereas grafting with chitosan or chitooligosaccharide reduces the biodegradability. In contrast to the chitosan-grafted-PHBV membrane, the chitoooligosaccharide-grafted-PHBV membrane showed lower antibacterial activity but higher biodegradability.[68]

Diazo linkage derived from diazo polyesters of ethylene glycol and ethylene glycol methacrylate were used as macro initiators in the process of forming different graft copolymers that were used in smart hydrogels for biomedical applications.[69–72] Previously, the elastic response of thermoplastic polyhydroxyalkanoates (PHOU) was improved upon cross-linking the side-chain olefinic bonds using peroxide as a cross-linker.[73,74] The study showed that the degree of cross-linking is a function of the type of peroxide and its concentration, and the cross-linking improved the elastic response by reducing the material crystallinity and tensile strength.[73,74]

7.3 Physical Modification of PHAs

The diverse potential applications of PHA in a number of fields demanded the production of smart polymers with minimal toxic impurities. Chemical modification methods are sometimes aggressive, and lead to reduced polymer molecular weight, unwanted side reaction(s) and toxic impurities. In some instances, a mild surface modification process is required without which the polymer may fail in its intended application(s). For example, neat polymer without the proper modification may cause delamination of adhesive bonds, poor cellular attachment, permanent staining of a fabric, or may influence proteinaceous membrane fouling *etc*.[75] These and many other reasons necessitate the application of physical methods (Table 7.1) in polymer modifications, as explained in the subsequent sections.

7.3.1 PHA Blending and Coating

Polymer surface roughness, chemistry and thermodynamic properties are among the influential features in polymer biocompatibility.[76] Studies have shown that the surface properties of polymers determine the type of molecules that can be adsorbed.[76–78] For example, in tissue engineering, ideal polymeric scaffolds should suitably provide microenvironments for cellular attachment, support, regeneration, proliferation, differentiation and tissue neo-genesis.[79] Cellular attachment and adhesion are suggested to be highly dependent on receptor-mediated interactions that rely on mutual molecular recognition between receptor integrins located on the cellular surface and ligands on the biomaterial surface.[79] Although diverse types of PHA could be used as tissue scaffolding, these materials are reported to lack ligands for biological recognition.[80] Hence, several physical approaches were devised in order to improve the cellular adhesion potential of PHA and its composite (Table 7.1). Blending PHBHHx with PHB resulted in a dramatic increase in the attached L929 cell viability as compared to the neat polymer or PHBHHx digested with lipase or NaOH.[81] On the other hand, coating PHBHHx with hyaluronan smoothed the polymer surface but decreased the contact angle of water to the material surface by approximately 30% resulting in about a 40% reduction in the growth of attached mouse fibroblast (L929) cells when compared with the neat polymer.[77] This kind of coating and blending could be employed in biomaterial design for cellular growth selection.

Similarly, blending the PHA with the anti-fouling agent 4,5-dichloro-2-*n*-octyl-4-isothiazolin-3-one (DCOI) improved the polymer's stability towards

Table 7.1   Typical PHA physical modification methods.

PHA	Modification method	Results/application	Ref.
PHBHHx	Hyaluronan coating	Improved hydrophilicity, low fibroblast L929 cellular growth 1.6×10^5 $(4 \times 10^5)^a$ cells per ml. Could be employed in biomaterial selection and design.	77
PHBHHx	PHB blending	Improved biocompatibility and higher fibroblast L929 cellular growth 1.9×10^5 $(1.8 \times 10^4)^a$ cells per cm^2. Could be employed in biomaterial selection and design.	81
mcl-PHA	Crosslinking by γ-irradiation	Higher cross linked density, decreased biodegradability. Increased Young's modulus 129% $(114\%),^a$ improved tensile strength 76% $(35\%).^a$	82
PHO, PHU	PEGMA grafting by UV irradiation	Reduced blood protein adsorption, platelet adhesion, and improved blood compatibility. Potential application in blood contacting devices.	83,84
PHB	OH$^-$ ion implantation	Improved bioactivity, wettability. Suitable for cell culture scaffold.	85
PHBV	O_2 plasma treatment	Increased hydropilicity, reduces polymer surface roughness, resulting in high cell growth and attachments. A potential scaffold for retinal pigment epithelium cell culture.	86
PHBV	Electrospinning	Improved elastic property, increased wettability. Scaffold for tissue engineering.	87

^{a}Numbers in parentheses represent unmodified PHA control values.

environmental degradation and reduced biofouling caused by soil microbes.[88] In fact, blending the PHA with DCOI was observed to postpone the onset of weight loss by up to 100 days, about a 10-fold increase in degradation stability compared to neat PHA film.[88]

Previous studies have shown that PHA granule binding protein (PhaP) and its fusion proteins, such as arginine-glycine-aspartate (RGD) tri-peptide, have higher cell-binding sequences, which improve cell survival, attachment, motility and proliferation, allowing their usage in specific drug delivery devices and cell sorting.[89,90] You et al.[89] compared the biocompatibility and cell viability of a scaffold made of neat PHBHHx to that of PHBHHx coated with PHA PhaP fused with RGD (PhaP-RGD). The study showed that the PhaP-RGD coating led to better homogeneity in cell spreading, adhesion, proliferation with improved chondrogenic differentiation and high extracellular matrix production compared with the neat PHBHHx scaffold.

Similarly, PHA repressor protein (PHaR) was reported to contain a hydrophilic DNA binding domain (DBD) and hydrophobic granule binding domain (GBD) allowing it to bind different polymers and biological macromolecules.[91] On the other hand, extracellular matrix proteins such as *Lys-Gln-Ala-Gly-Asp-Val* (KQAGDV) oligo-peptide have been shown to enhance cell adhesion and proliferation.[92] Recently, Dong *et al.*[79] evaluated the cytocompatibility of PHA coated with PhaR and *Lys-Gln-Ala-Gly-Asp-Val* (KQAGDV) polypeptide. Using water contact angle measurements, the coated polymer exhibited a significant improvement in hydrophilicity. Furthermore, in terms of cytocompatibility, the PhaR-KQAGDV coating clearly enhanced the adhesion and proliferation of human vascular smooth muscle cells (HvSMCs) far better than the non-coated controls based on microscopic investigations and CCK-8 assay.[79]

7.3.2 PHA Irradiation

In contrast to other polymer modification methods, irradiation of polymeric materials required no addition of polymer contaminants. Irradiations such as gamma-irradiation normally result in three-dimensional network structures with improved tensile strength. Several studies have demonstrated the cross-linking of unsaturated mcl-PHAs by gamma-irradiation.[82,93,94] The presence of olefinic bonds in PHA side chains provides an avenue for polymer modification by several irradiation processes. A highly cross-linked modified polymer was produced by irradiating unsaturated PHA obtained from tallow-grown *P. resinovorans* with 25–50 kilogray (kGy) of γ-irradiation. The modified polymer showed a reduced degradability, increased Young's modulus and enhanced tensile strength (Table 7.1).[82] The study also showed that radiation treatment up to 50 kGy had no effect on the thermal stability, melting enthalpy (ΔH_m) and storage modulus of the polymer, but induced polymer chain scission.[82] The polymer chain scission was minimized by the addition of linseed oil, which caused a 2.5-fold reduction of ΔH_m in the entire polymer tested. This observation was attributed to the high density of olefinic bonds in linseed oil.[82] A modified cross-linked polymer of PHOU was also prepared by γ-irradiation.[94] Optimal cross-linking was found upon treatment with a 20 kGy dose in an inert atmosphere (N_2). The polymer film was reported to be unaffected upon mild radiation treatment.[94] Smart hydrogels of semi-interpenetrating polymer networks (IPNs) based on PHB and a hydrophilic poly(ethyleneglycol diacrylate) monomers at different compositions were prepared by radiation-induced polymerization using γ-rays within 10–100 kGy.[95] A reduction in polymer thermal melting temperature and crystallinity with increasing radiation dose was observed. This was probably due to the partial destruction of the crystalline region upon exposure to the radiation energy.[96] Simultaneous irradiation techniques were reported to produce a homogeneously grafted polymer of high purity.[97] The effect of solubilizing solvent on the simultaneous γ-irradiation induced graft copolymerization of acrylamide onto PHB was also studied.[97] The authors use three different models (eqns (7.1), (7.2) and (7.3)) to describe the percentages of grafting degree ($W\%$), monomer/macromer composition ($C_x\%$) and crystallinity index (CI%), respectively.

$$W\% = \frac{m_g - m_i}{m_i} \times 100 \tag{7.1}$$

$$C_x\% = \frac{m_x}{m_g} \times 100 \tag{7.2}$$

$$CI\% = \frac{A_{\lambda CH_3}}{A_{\lambda C-O-C}} \tag{7.3}$$

where m_g, m_i and m_x are thermogravimetric analysis (TGA) derived graft copolymer weight, neat backbone polymer weight, and weight of the monomer or macromer component under analysis, respectively. $A_{\lambda CH3}$ is the FTIR integral peak area of the band at wavelength (λ) 1382 cm^{-1} assigned to methyl bending vibrations, which is insensitive to the crystallinity changes, and $A_{\lambda C-O-C}$ is the integral peak area at 1185 cm^{-1} corresponding to ester vibrations that is sensitive to changes in crystallinity.[97] In all the solvents evaluated (acetone, methanol, ethanol, ethyl acetate and chloroform), the researchers found that the sample grafted in chloroform had the highest degree of grafting ($\approx 91\%$) in spite of the chloride radical that was reported to terminate the reaction.[97] In addition, the chloroform-grafted sample was found to exhibit higher water uptake, about 2.5-fold higher than the sample grafted in acetone. This could be due to an increased degree of acrylamide grafting as a result of higher polymer solubilization in chloroform.[97] Several PHA modifications by gamma induced graft copolymerization have been reviewed elsewhere.[98]

Olefinic bonds of unsaturated polymers confer a cross-linking/grafting advantage upon irradiation with ultraviolet (UV) light even in the absence of photosensitizers or initiators.[84] On the other hand, saturated polymers such as PHO normally need photo-initiators (*e.g.* benzophenone) to initiate the cross-linking reaction upon UV irradiation.[83] A series of unsaturated PHAs derived from linseed oil was cross-linked using UV radiation at a wavelength of 300 nm, which resulted in modified PHAs with increased glass transition temperatures and improved curing due to the cross-linking reactions.[99]

For biomedical applications, inhibition of protein adsorption, as well as platelet activation and adhesion on the polymer surface, is critical to the efficiency of the material. For example, use of PHAs as artificial blood con-tacting devices such as arteries and anticoagulant films was limited by surface-induced thrombosis.[83,100] The adsorption of plasma proteins and adherence of activated platelets onto the polymer surface resulted in their transformation to pseudopods and subsequent release of platelet biochemical content, which in turn activated other platelets leading to the thrombosis condition.[83] The monoacrylate of polyethylene glycol metha-crylate (PEGMA) was grafted with poly(3-hydroxyundecenoate) (PHU)[84] and PHO[83] by UV irradiation of the homogeneous PHA solutions containing PEGMA. In both studies, the surface adsorption tendencies of both graft copolymers towards blood proteins and platelet adhesion were reduced significantly in comparison to the control *i.e.* poly (*L*-lactide) surface. Fur-thermore, the grafted copolymers exhibited excellent blood compatibility with increasing PEGMA fraction.[83,84] A higher PHA molecular weight (M_w) is associated with a lower degradation rate thus limiting its application as a short-lived drug carrier for molecules such as DNA and functionalized polypeptides. Instead of cross-linking and grafting the PHA, Shangguan *et al.*[101] used UV irradiation to achieve controlled degradation of PHBHHx leading to oligomers bearing reactive radical groups with a quick degrad-ation rate. The authors found the decrement of the M_w to be dependent upon the UV radiation exposure time. The degradation mechanism was extensively reviewed.[9]

7.3.3 Ion Implantation

Ion implantation is another physical method employed in polymer surface modification. Its advantage over other polymer modification methods is that it only modifies the polymer surface layer, without upsetting the bulk polymer's properties.[102] Ion implantation has been successfully applied in several polymer modifications thereby expanding its applications.[102–106] Hou *et al.*[85] successfully implanted hydroxyl ions on the surface of PHB using an electron ion implantation energy of 40 keV when an implantation flux ranging from 1×10^{12} to 1×10^{15} ions per cm^2 was applied. The study reported that the modified PHB showed a better proliferative activity of mouse embryo fibroblast (3T6) cells compared to non-modified PHB. However, it was observed that the ion implantation caused polymer cracks proportional to the intensity of fluence or flux of the hydroxyl ion bom-bardment.[85] In a similar study, Santos *et al.*[107] improved the hydrophilicity of PHB and PHB-graft-polyvinylacetate copolymers by implanting H^+, Ag^+, and Na^+ ions at different fluences. In both polymers, the treatment re-sulted in increased hydrophilic characteristics. PHB and PHBHHx wett-ability was shown to be significantly improved upon implantation with carboxyl ions at an electron implantation energy of 150 keV and ion bombarding fluences ranging from 5×10^{12} to 1×10^{15} ions per cm^2.[102] In comparison to the non-implanted PHA, the polymer surface analysis re-vealed a decrease in the intensities of $-C-C-$, $-C-O$ and $-C=O$ groups in the COO^- implanted polymer, thus improving its hydrophilicity. Likewise, PHB, PHBV and PHBHHx implanted with C^+ ions revealed im-proved cytocompatibility when used as supports for culturing mouse em-bryo fibroblast (3T6) cells.[105] In all polymer samples, it was found that an implantation ion flux of 1×10^{12} ions per cm^2 was the optimal in conferring better cytocompatibility.

7.3.4 Plasma Treatment

When a gas is subjected to extreme heat, or an electric or electromagnetic field applied from a laser or microwave generator, this causes the molecules to ionize thus turning the gas into plasma composed of charged ions, electrons, radicals, and neutral species.[108] When the generated plasma has a uniform thermal equilibrium with its components at the same temperature, such plasma is described as thermal plasma. On the other hand, the plasma is termed a non-thermal having a strongly deflected kinetic equilibrium when excited electrons temperature is higher than the ions and the neutrals.[109] Because of their high temperature, diverse density and complex compositions, plasma interacts with a polymer when it is physically bombarded by excited electrons resulting in polymer surface modification bearing chemical or ionic groups that could participate in further reactions such as cross-linking, grafting, etching, roughening and functionalization. Like ion implantation, plasma treatment has the advantage of modifying the material's surface layer only, without tempering with the polymer's intrinsic mechanical properties.[110] The high discharging efficiency of radio frequency (RF) plasma makes it an excellent choice in this regard.[108]

Despite its excellent biodegradability, compatibility and diverse adjustable mechanical properties,[7] the highly crystalline PHB is strongly hydrophobic, resulting in limited applications due to its slow degradation rate, biorsorbability and poor cell adhesion.[26,28] Surface treatment by plasma is considered as an effective approach to increase the hydrophilicity and wettability of such polymers.[111] Recently, Mirmohammadi *et al.*[108] compared the biocompatibility of the PHB surface upon treatment with O_2 and CO_2 plasma at 50 W discharge for 3 min, and found that O_2 plasma treated PHB showed much improvement. The results indicated that neat PHB with an irregular and coralloid surface was modified to a more regular surface morphology with improved roughness (nano-protrusion and nano-indentation) upon plasma treatment. This modification resulted in enhanced cell–polymer electrostatic interactions, and improved growth of attached L929 fibroblasts. Based on FTIR-ATR analysis, it was observed that the surfaces of plasma treated PHB were functionalized *via* endowment of oxygen functional groups such as –OH, COO– and –CO– that improved the wettability of the treated surfaces.[108] Hasirci *et al.*[112] studied the influence of oxygen RF-plasma treatment on the surface and bulk properties of PHBV. Their findings showed that the plasma-treated films absorbed more water than the untreated films, and the degree of absorption depends on the applied plasma discharge power. They further observed a decrease in water contact angles and an increase in oxygen–carbon atomic ratio upon treatment, indicating improved hydrophilicity due to an increase in the oxygen-containing functional groups on the surface of the polymer.[112]

Radio frequency glow discharge (RFGD) plasma generated at 100 kHz was used to modify a PHO surface that was subsequently grafted with acrylamide in aqueous solution.[110] The amount of grafted amide and wettability were found to increase proportionally to the increasing plasma discharge power. The modified polymer biocompatibility was evaluated using Chinese hamster ovary cells where surfaces treated with 30 W of plasma have moderate hydrophilicity while showing excellent cell adhesion and high growth. Improved hydrophilic properties and highly stable modified PHB-amino groups were obtained when low pressure microwave ammonia plasma treatment of the polymer surface was used.[113] The presence of the amino group on the PHB surface was established using X-ray photoelectron spectroscopic (XPS) analysis after the treated sample was chemically labeled with 4-trifluoromethyl benzaldehyde.

As previously mentioned, the blood coagulation process (thrombosis) is a complex process that involves the participation of blood proteins and platelets. Protein adsorption onto polymeric material induced platelet adhesion leading to thrombosis. Thus, an effective anticoagulant film is expected to have high protein resistance characteristics. Increased surface wettability, charge, morphology and functional groups are among the factors indicated to help inhibit protein adsorption onto a polymer's surface.[114,115] The wettability and protein resistance characteristics of P3HB4HB were improved upon treatment with oxygen plasma for 10 minutes.[116] The study showed that the platelet resistance characteristic of the polymer was improved by 96.8%; likewise, the protein (BSA) resistance characteristic was also improved by 27% and 57.5% in phosphate buffer and aqueous solutions, respectively.

It has been reported that an effective cure for retinal disorder due to retinal pigment epithelium (RPE) degeneration is currently unavailable.[86] Oxygen plasma treatment was used to improve the hydrophilicity of a scaffold based on PHBV containing 8 mol% hydroxyvaleric acid, which was used to successfully culture RPE (D407) cell lines. An engineered scaffold with a rougher surface and enhanced hydrophilicity for cardiovascular tissue was prepared from PHBHHx-coated silk fibroins modified using low temperature atmospheric plasma.[117] Similarly, ammonia plasma treated PHBHHx followed by fibronectin coating encouraged better growth of human umbilical vein endothelial cells (HUVECs) and rabbit aorta smooth muscle cells (SMCs) on its surface compared to the untreated polymer, and could serve as a potential material for the luminal surface of vascular grafts.[118] A plasma-induced polymerization technique using polyethylene glycol (PEG) and ethylenediamine (EDA) as a co-monomer was used to modify the surface of an electrospun PHB nanofibrous scaffold resulting in a material with enhanced cellular proliferative activity.[119] Similarly, the same plasma-induced polymerization was used to graft polyacrylic acid onto PHAs of different monomeric compositions, resulting in modified PHAs with relatively unchanged mechanical properties but enhanced hydrophilic properties.[120]

7.3.5 Electrospinning

Electrospinning is a method that employs the use of a high electrical voltage to produce polymer fibers of different diameters from polymer

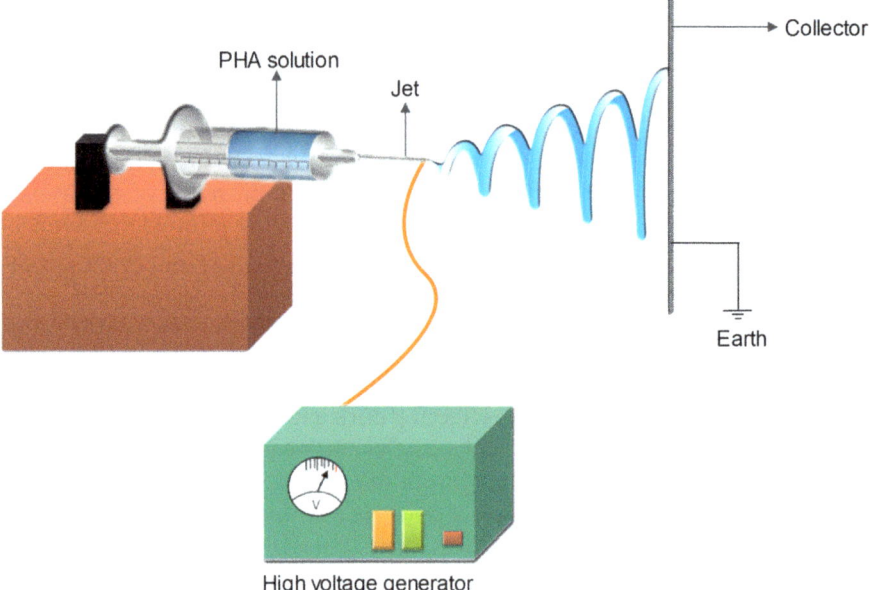

Figure 7.11   PHA electrospinning set-up.

solutions (Figure 7.11).[121,122] Polymer electrospinning is used to produce a polymeric scaffold that mimics natural extracellular matrices. The solution's viscoelasticity, its surface tension and jet charge density were reported to be among the influential factors in polymer electrospinning processes.[122] Jet charge density is affected by the applied electrostatic field and solution conductivity. For example, Fong *et al.*[122] reported that addition of sodium chloride to an aqueous solution of polyethylene oxide increased the net charge density of the spinning jet. In the case of a low molecular weight polymer solution, the researchers reported that the electrically driven spinning of the solution resulted in polymer droplets (electrospray) due to the capillary breakup of the spinning jet as a result of surface tension.

Kwon *et al.*[87] used a high voltage (10–20 kV) to produce ultrafine electrospun PHBV fiber mats that were used as scaffolds for tissue engineering. The researchers observed that the electrospun nanofiber's diameter decreased with the decreasing concentration of polymer solution and increasing applied voltage. In comparison to the neat polymer, they found that the electrospun PHBV exhibited about a six-fold increase in elongation to break than the PHBV cast film. The study showed that due to the increased porosity and surface area of the electrospun PHBV fibrous mat, seeded chondrocytes adhered to and grew better on it than on the cast PHBV film. A similar study on the improved biocompatibility of electrospun PHBV was reported.[123] When studying the cell attachment efficiency of the electrospun nanofibers by seeding chondrocytes derived from rabbit ears on both the PHBV cast film and electrospun PHBV nanofibrous mat, the researchers observed about a two-fold increase in the chondrocytes' attachment after two hours of incubation when electrospun nanofiber was used as compared to the cast film.[123] Yu *et al.*[121] observed a decrease in the crystallinity and crystallization rate by doping zinc oxide (ZnO)/PHBV electrospun nanofiber with ZnO nanoparticles. They attributed the observed effect to the interaction between the hydroxyl groups on the ZnO nanoparticles' surface and the polymeric carboxylic groups. Ying *et al.*[124] evaluated the biocompatibility and biosorption characteristics of the electrospun scaffold of P3HB4HB through subcutaneous implantation of the fibers in rats. The researchers found a highly increased tissue response with increasing content of 4HB monomer.

7.4 Modification of PHAs with Enzymes

PHA modification *via* an enzyme-mediated process is seen as a mild, specific and environmentally-friendly method. In this section, PHA modification using enzymatic degradation and/or synthesis methods in both *in vivo* and *in vitro* processes is discussed. Also included in the discussion is enzymatic modification of PHA using the degradation products of PHA itself.

7.4.1 *In Vivo* Enzymatic Degradation of PHA

During the bacterial fermentation process, PHA granules are accumulated inside the cells in the presence of excess carbon source(s) but limited essential nutrients such as nitrogen, oxygen, phosphorous, potassium, or sulfur.[125] In a reverse situation where the microbes are starved of a carbon source in the presence of abundant nutrients, they will start producing intracellular PHA depolymerase and dimer hydrolase to degrade the accumulated PHA and continue growing.[126] The monomer produced from the intracellular degradation is subsequently oxidized by the cell to form acetoacetate. In an attempt to bypass the accumulated PHA hydrolysis, Lee *et al.*[127] proposed a process that incorporates continuous limitation of the carbon source and nutrient(s) in an anaerobic condition. The absence of oxygen inhibits the cells from metabolizing the 3HB monomers from intracellular PHB degradation by lowering the concentration of *R*-3-hydroxybutyrate dehydrogenase, which in turn minimizes the conversion of 3HB to acetoacetate.

It has been reported that a high yield of 3HB (96%) can be obtained within a relatively short time (30 min) in *Alcaligenes latus* by using a fed-batch culture system with sucrose as a carbon source. The cells with stored PHB were collected and incubated at pH 4 and 37 °C to provide an environmental condition in which cells exhibit high activity of intracellular PHA depolymerase and low activity of (*R*)-(−)-3-hydroxybutyric acid dehydrogenase.[127] Other identified factors that contribute in *in vivo* depolymerization are substrate concentration and extracellular pH. Ren *et al.*[128] reported that the optimal initial pH range for initiation of intracellular depolymerization of PHA by *Pseudomonas putida* was 8–11, and pH 11 after commencing monomer release.[128] A drop in the solution's pH was due to secretion of PHA monomers by the bacteria.[12]

7.4.2 *In Vitro* Enzymatic Depolymerization

7.4.2.1 *Extracellular PHA Depolymerase*

Many PHA-degrading microorganisms have been isolated and their ability to secrete extracellular PHA depolymerase could be exploited. The degraded monomer can be extracted and used as a macro-initiator for further modification.[129] The extracellular PHA depolymerases for mcl-PHA have been identified mainly from Gram-negative bacteria, predominantly *Pseudomonas* species; *Pseudomonas fluorescens* GK13, actinomycete species; *Streptomyces roseplus* SL3, and scl-PHA depolymerase from *Alcaligenes faecalis* T1.[130–132] This enzyme belongs to the family of serine hydrolases with lipase consensus sequence *Gly-X-Ser-X-Gly* and it is strongly hydrophobic.[133] The catalytic triad for PHA depolymerase consists of a serine residue that acts as a nucleophile with aspartate (or glutamate) and histidine to stabilize it (serine-histidine-aspartate).[134,135]

Shirakura *et al.*[131] observed that PHB depolymerase acts as an endo-type hydrolase as the enzyme only cleaves at the second ester linkage from the hydroxy terminus of the trimer, tetramer, pentamer and higher oligomers as shown in Figure 7.12. At the same time, the PHA depolymerase prefers PHA with alkyl side chains rather than linear PHA.[135] According to Mukai *et al.*,[136] PHA depolymerase enzyme is very selective towards chiral monomers, especially alkyl side chains such as 3HB rather than 4HB. The rate of enzymatic hydrolysis is primarily dependent on the composition and length of the PHA side chain. Longer side chains provide steric hindrance for the enzyme to be adsorbed effectively on the polymer backbone chain as shown in Figure 7.13.[137]

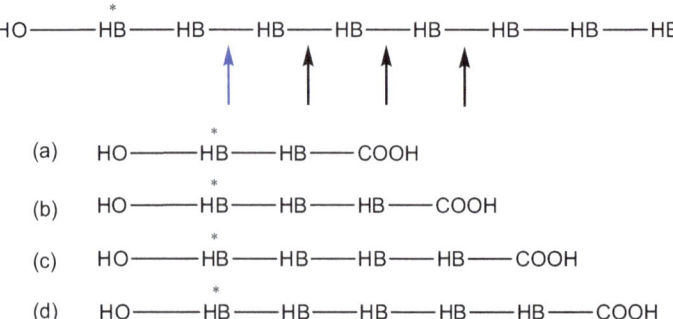

Figure 7.12   Bond cleavage during initial action of poly(3-hydroxybutyrate) depolymerase on radioactive oligomers start from blue arrow. (a), (b), (c), (d) are some products of hydrolysis with (a) degradation product not undergoing further hydrolysis.

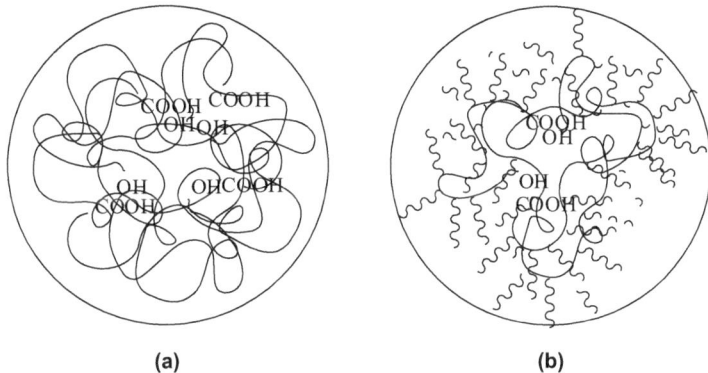

Figure 7.13 Schematic folding structure of (a) linear PHA, (b) branched PHA.

Several key factors in enzymatic catalysis include their activity, stability, substrate concentration and temperature. Optimization of these factors will steer the enzyme towards the desired activity. Media pH was shown to affect the ionizable groups at the active site of the enzyme.[138] Gangoiti *et al.*[139] reported that P(3HO) depolymerase showed its maximum activity at pH 9.5 and a temperature of 40 °C.

7.4.2.2 Lipases

Lipases are subclass of esterases, capable of hydrolyzing esters, fats and lipids in aqueous media, hence they are described as hydrolases belonging to class EC 3.1.1.3. Lipases are one of the most versatile enzymes because they can catalyze many different reactions such as esterification, interesterification, hydrolysis, alcoholysis, peroxidation, aminolysis, and epoxidation.[140,141] Most lipases have similarities in their amino acid sequence including within the catalytic region, ***His-X-Y-Gly-Z-Ser-W-Gly*** or *Y-Gly-**His-Ser**-W-Gly* where *W, X, Y and Z* refer to unspecified amino acid residues.[142] These enzymes are stereospecific towards ester bonds thereby eliminating any undesirable by-products of the reaction.[143] Lipase has been used extensively as a biocatalyst in industry because of its high thermal stability, versatile pH range and it can be used repeatedly if immobilized (*e.g.* Novozyme 435).[144]

In PHA modification, Mukai *et al.*[136] reported that lipases originated from eukaryotes have broad specificities with the abilities to erode P(3HP), P(4HB), P(5HV) and P(6HH) films compared to prokaryote lipases, which could barely degrade all the polymers except P(3HP). This shows that lipases from prokaryotes have high substrate specificities for the hydrolysis of PHA. Jaeger *et al.*[135] indicated that *Pseudomonas alcaligenes, Pseudomonas fluorescens, Pseudomonas aeruginosa, Burkholderia lemoignei* and *Bacillus subtilis* lipases show specificities towards ω-hydroxyalkanoic acid over α-hydroxyalkanoic acid. The absence of alkyl side chains in the polymer backbone allows for more flexibility and hydrophilicity of the aliphatic main polymer chain thus resulting in better engagement between the polyester chain and the active site of lipases.[136,145,146] For lipase catalyzed degradation and esterification reactions, different factors require special attention. In order to optimize lipase catalyzed esterification activity in organic solvents, the temperature of the reaction plays an important role in enzyme stability and the solubility of substrates such as sugar moieties and alcohol since the major problem with hydrophilic moieties is that they are insoluble in organic solvents.[147–149] It has been reported that at a temperature of 50 °C, a high yield of poly(1'-*O*-3-hydroxyacyl-sucrose) formed using Novozyme 435 was shown to be due to an improvement in substrate dissolution.[28] While for lipase catalyzed PHA degradation in toluene, the reaction rate was almost the same between 40 and 60 °C.[150] However, factors such as type of lipase,[135] concentration[146] and degree of initial crystallinity[151] were reported to influence the rate of PHA degradation by lipases.[151]

7.4.3   Degradation Products

7.4.3.1   Chiral Monomer of PHA

Recently, there has been a surge in demand for pure biodegradable PHA enantiomers due to their reported biological activities.[152] The extensive use of pure enantiomers in industries such as medicine, agriculture and the food industry boosts the need for its large scale production.

(*R*)-(–)-Hydroxycarboxylic acid can be widely used as a chiral building block for the synthesis of fine chemicals such as antibiotics, vitamins, aromatics, and pheromones.[153] Its hydroxyl and carboxylic groups are amendable to modification in addition to being utilized as precursors for the synthesis of new compounds.

In another study, PHA synthase from *R. eutropha* was used to polymerize PHB and PHV on hydrophobic highly oriented pyrolytic graphite (HOPG) and alkanethiol self-assembled monolayer (SAM) surfaces, which are used to support other functional biomolecules such as streptavidin and biotin. The surface modification was reported to have biomedical and biotechnological applications.[154]

7.4.3.2   PHA Oligomers

There has been a great deal of interest in PHA oligomers because of their *in vivo* biodegradability and bioresorbability. Furthermore, Tasaki *et al.*[155] stated that dimers and trimers of 3HB can be rapidly converted to monomers in rat and human tissues. Thus, various kinds of dendrimers made up of oligo-HA could be developed and used for various drug delivery applications.[152] Oligomers from the degradation of neat PHA can be used for grafting copolymers at a carboxylic end terminus. Oligomers show a narrower molecular weight distribution compared with neat PHA and they come with relatively higher functionality. Grafting oligomers allows more controlled processes and a high yield (83%).[156]

7.4.4 Surface Modification

PHAs' excellent mechanical properties, biocompatibility and degradability have made them useful in the field of tissue engineering. These polymers could prove very useful as tissue scaffolds for implantation purposes.[157] However, the smooth surface of a solvent-cast PHA scaffold is a major obstacle for cell attachment in tissue regeneration processes.[158] This warrants the enzymatic catalyzed surface erosion of PHA. Moreover, the PHA surface lacks a bioactive ligand to couple with a bioactive molecule in targeting devices or biosensors. Hence, PHA surface erosion or roughening is needed to provide a corrugated trough to immobilize bioactive molecules such as insulin,[159] fibronectin[160] and collagen[161] while enhancing its cell attachment or cell proliferation characteristics and thus expanding its biomedical applications.[162] Ihssen *et al.*[163] used extracellular PHA depolymerase from *Pseudomonas fluorescens* as the capture ligand to immobilize a fusion protein on to mcl-PHA microbeads of 200–300 nm size (Figure 7.14). This arrangement could be employed as a probe for targeting proteins in drug delivery, protein microarrays, and protein purification. Furthermore, the researchers reported that the binding capacity was comparable to similar-sized polystyrene particles commonly used for antibody immobilization in clinical diagnostics.[163]

Recently, the use of nanoparticles as drug delivery systems for targeted release at specific parts of the body represents a promising solution for cancer treatment.[164] Surface functionalization of PHA nanoparticles is needed to improve targeting efficiency in delivering drug molecules to target cells. This modification can be done *via* PHA-protein block copolymerization. Generally, during *in vivo* synthesis of PHA, PHA synthase is a key enzyme that catalyzes the polymerization of hydroxyacyl-Coenzyme A, a

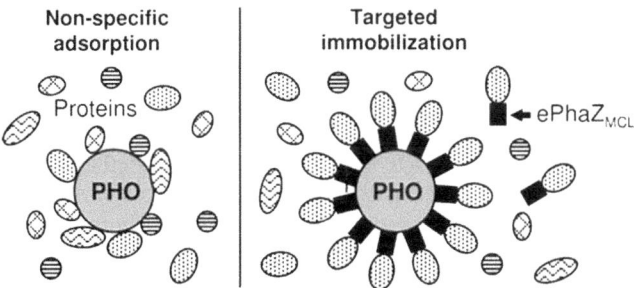

Figure 7.14 Immobilization of targeted protein onto PHA using extracellular PHA depolymerase, reprinted from Ihssen *et al.*[163] with permission from the American Chemical Society.

substrate for PHA biosynthesis. On the other hand, during *in vitro* PHA synthesis, PHA synthase and the hydrophobic chain of the polymer were covalently bound resulting in the formation of an amphiphilic block co-polymer in which PHA synthase can further be modified through protein engineering to improve the protein-PHA copolymer.[165,166] It has been re-ported that the first 100 N-terminal amino acid residues of PHA synthase can be removed without affecting the enzyme's activity.[167] Thus, additional lig-and fusion can still be carried out without affecting the protein's native catalytic activity. Incorporating a tumor specific ligand, RGD4C, with PHA synthase helps PHA nanoparticles adhere more effectively on MDA-MB 231 breast cancer cells.[168] Modifications of the PHA nanoparticles were de-scribed by Kim *et al.*[165] where the use of PHA synthase from *Ralstonia eutropha* H16 for the formation of protein-PHB copolymer micelles (involving polymerization and self-assembly of 3-hydroxybutyryl-CoA (3HB-CoA)) in aqueous solution at room temperature resulted in a hydrated shell. Concomitantly, hydrophobic drug molecules are integrated into the core of the shell. On the other hand, Lee *et al.*[168] proposed a coupling between PHB emulsion and enzymatic protein functionalization. Then growing PHB tail chain arising from the action of PHA synthase that was fused earlier with RGD4C and this arrangement was able to form hydrophobic interactions with the surface of the PHB nanoparticle emulsion containing drug mol-ecules. As the tail chains grew continuously, the surface of the nanoparticles was covered with RGD4C peptide-PHB copolymer that targets breast cancer cells. Paik *et al.*[166] showed that PHB nanoparticles can also bind to a solid surface. In their study, PHA synthase was fused to a *His*-tag (10x-histidine) expressed in a recombinant *E. coli* synthesizing PHB to produce a protein–polymer hybrid with *His*-tag end-functionality. The *His*-tag bound tightly with a Ni^{2+}-nitrilotriacetic acid (Ni-NTA) derivatized solid surface (silicone or agarose). All these modification methods are excellent demonstrations of synthesizing a wide variety of protein functionalized PHAs with novel properties.

PHA surface erosion is normally initiated at micro-holes on the PHA's surface that allow enzymes and water molecules to adhere to the film sur-face, commencing the hydrolytic process.[169,170] At the beginning, water molecules enter the amorphous regions within the film, which triggers the enzyme-catalyzed hydrolysis of the ester bond.[171] During the hydrolysis process, the roughness of the PHA film increases over time.[172] While the enzyme is thought to attack mainly the amorphous region, wide angle X-ray diffractography revealed a decline in the crystalline peak after 22 hours of reaction, which indicated that the crystalline region was also hydrolyzed by lipase after the amorphous region was eroded.[172] Selective surface erosion using PHB depolymerase helps to liberate the 3HB monomers from the surface of P(3HB-*co*-4HB) and leave the undegradable 4HB monomers on the polymer surface.[173] Different polymer constitutions on the polymer surface may exhibit different degradation properties. Among the parameters that contribute to the enzymatic degradation of PHA films are the molecular weight of the polymer, crystallinity, monomer composition, porosity and roughness of the polymer surface.[173] In order to measure the rate of degradation, changes in the weight loss of the PHA film over time can be monitored. While the biopolymer chains could be completely degraded into monomers and dimers of 3-hydroxyalkanoic acids by PHA depolymerase,[174] this process is very slow. Williams *et al.*[175] measured the molecular weights at the surface *versus* the interior of the PHO implants but no significant differences were detected. This showed that homogeneous hydrolytic breakdown of the polymer occurred inside the mice.[176]

7.4.5 PHA Functionalization

Previously, applications of lipase had been limited because the enzyme was thought to work effectively only in environments with a high water content, and would rapidly lose its activity in organic solvents.[177] Later, the enzyme was found to remain catalytically active in organic solvents, and this generated huge interest among researchers and industries as most of the industrial substrates are hydrophobic in nature. Utilization of an organic solvent as a reaction medium helps to improve hydrophobic substrate solubility and simultaneously elevates the rate of the reaction. In addition, a micro-aqueous environment enables reactions that are impossible in water, such as esterification, to be carried out.[178,179] It has been suggested that organic solvents may play a role in prolonging enzyme activity by replacing the molecules of water on the enzyme with solvent molecules.[177] Reactions in water usually caused downstream difficulties in the separation of soluble enzyme and products. However, in organic solvent media, the enzyme is insoluble and this facilitates product recovery.[180]

The functionalization of polymers may involve chemo-enzymatic synthetic steps. This can be seen in the esterification of poly (ε-caprolactone) with hydrophilic moieties in the presence of CAL B as a biocatalyst where tetrahydrofuran, dichloromethane and dioxane were used as reaction media.[149,181] Longer chains of PHA pose a steric hindrance for the enzyme to bind effectively with the polymer substrate at the carboxyl end terminus for transesterification. To overcome this limitation, enzyme catalyzed hydrolysis of the poly (ε-caprolactone) was carried out prior to chemical transesterification.[181] The enzymatic step not only facilitates the subsequent step but also helps to generate more carboxyl end terminals for transesterification.[81] In another study, Gumel *et al.*[182] studied the use of two organic solvents *viz.* dimethylsulfoxide (DMSO) and chloroform at a 1 : 4 ratio to dissolve reaction substrates with contrasting solubilities *i.e.* sucrose and PHA. Since a small amount of DMSO was sufficient to dissolve enough sucrose, and then mix with PHA in chloroform, a system that behaves largely like a single-phase system could be achieved. Consequently, the effects of interfacial resistance can be avoided. In this system, the researchers demonstrated a successful functionalization of a medium-chain-length PHA with sucrose thereby improving the modified polymer film's biodegradability.

Ravenelle and Marchessault[183] reported the transesterification of PHB with monomethoxy poly (ethylene glycol), mPEG using a bis(2-ethylhexanoate) tin catalyst at 190 °C to form diblock copolymers. However, they were unable to control the molecular weight of the PHB due to thermal degradation that occurs during the high temperature reaction. Gumel *et al.*[182] demonstrated the use of Novozyme 435 for the transesterification of PHA and sucrose in an organic solvent mixture at a mild temperature of 50 °C. From the GPC analysis, the polydispersity index showed a low value at 0.7 with a higher number average molecular weight (M_n) compared with a low weight average molecular weight (M_w). This indicated that less variation in the average molecular weight could be achieved using an enzyme-mediated transesterification process. The scheme of the transesterification process is shown in Figure 7.15.

In another study, Novozyme 435 was used to produce a novel PHB block copolymer based on PHB-PCL in microaqueous media.[184] Such block copolymers were reported to have expanded biomedical applications due to their excellent thermoplastic properties.[184]

The main factor affecting enzyme activity is the molecular water layer on the surface of the enzyme. The major causes of low activity are reduced conformational stability, uncontrolled pH and unfavorable substrate desolvation.[177] Since the amount of water retained naturally by the enzymes becomes the main factor affecting enzyme stability, solvent hydrophobicity (log P) value is a convenient indicator of solvent suitability as a reaction medium for the enzyme's activity.[185] While there is great concern over the accumulation of the water by-product that reverses the catalysis direction in lipase catalyzed esterification reactions, it has no relevance to other systems that utilize enzymes such as PHA synthase. For example, two new functionalized PHAs containing cyclopropane and chlorine *viz.* 3-hydroxy-3-cyclopropylpropionate (3CyP3HP) and 3-hydroxy-4-chlorobutyrate (4Cl3HB), respectively, were enzymatically produced using PHA synthase from *Escherichia shaposhnikovii* in an aqueous solution.[186]

At room temperature, PHA is soluble in chlorinated and other organic solvents such as dichloromethane, chloroform, carbon tetrachloride, dichloroethane, chloropropane, tetrahydrofuran, 1,2-propylene carbonate, toluene and hot acetone.[187] In lipase-catalyzed modification of PHAs, the benefits of using hydrophobic organic solvents over an aqueous system include increased solubility of nonpolar substrates, the enzyme favoring the ester-bond synthesis rather than hydrolysis, and elimination of microbial growth that usually contaminates the aqueous reaction mixture. Several solvent systems have been studied such as a water-and-hydrophilic solvent (monophasic), hydrophobic organic solvent (monophasic), water-and-water immiscible (two phase) solvent, and a nearly dry organic solvent system.[188] The advantage of a monophasic system is the minimal diffusion resistance between the substrate in the hydrophilic solvent and the water phase, and the maximum dissolution of substrate concentration that leads to an increase in the reaction rate.[188] In crystalline PHA, the polymer backbone

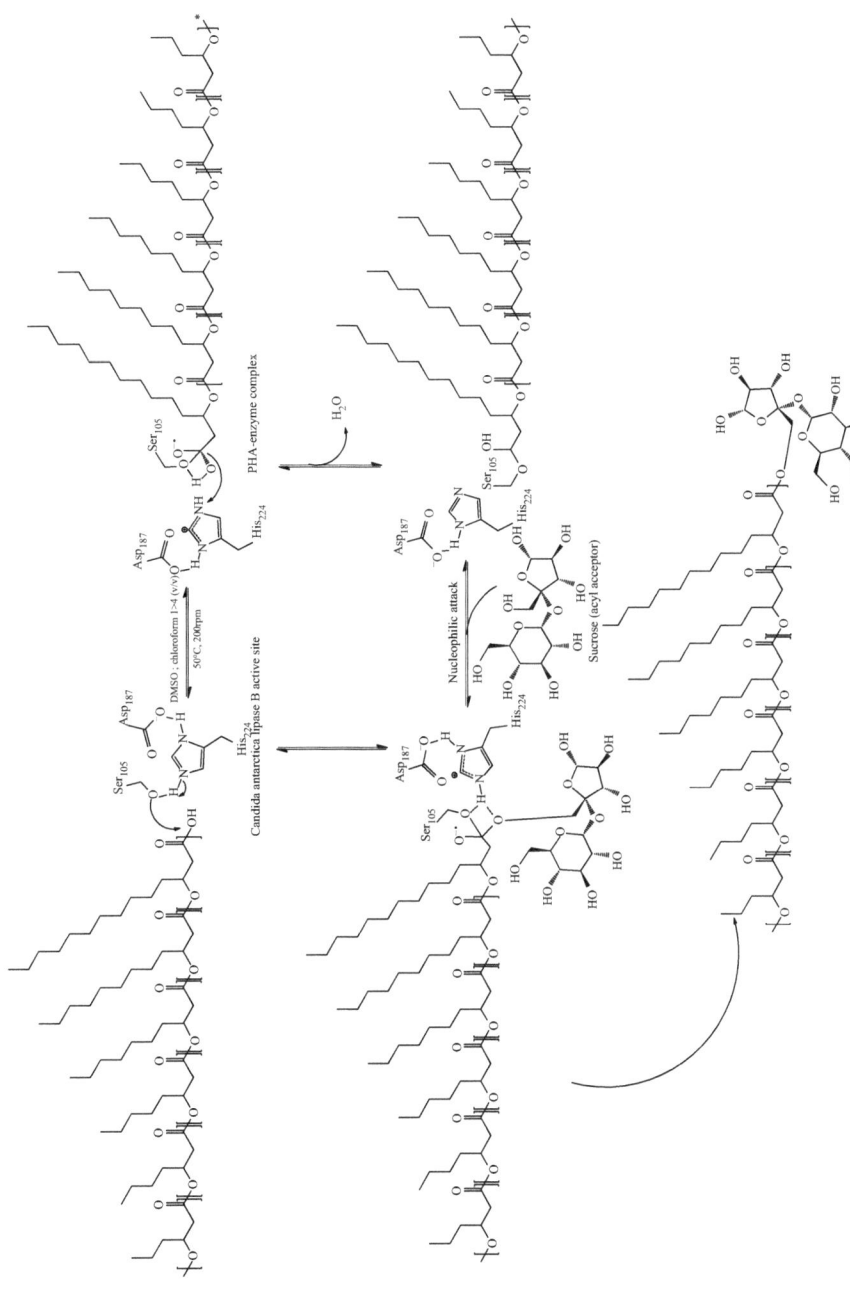

Figure 7.15 Scheme of transesterification of medium-chain-length PHA with sucrose using *C. antarctica* lipase B in an organic solvent mixture.[182]

chains were clumped together, leaving the alkyl side chains pointing outward, covering the polymer surface. When the polymers were dissolved in a solvent, the solvent molecules diffused through the polymer matrix until they reached the polymer core and stretched the polymer backbones to form a swollen, solvated mass. Then, it broke up and the polymer chains started to disperse into true solution. At this moment, the polymer backbones and both functional groups were exposed for enzyme attack.[189] Direct contact of enzyme with organic solvent results in the preferential partitioning of the enzyme's hydrophobic part towards the solvent with simultaneous changes in enzyme conformation, and this may eventually lead to enzyme deactivation.[190] When a two-phase system is introduced to minimize the inactivation of enzyme molecules, denaturing at the interphase between water and the organic phase may still occur albeit at a lower frequency.[188]

7.5   Conclusions

The demand for smart biodegradable PHAs with the flexibility to incorporate specific traits that extend their niche applications, and the difficulty encountered in their *in vivo* production by conventional biosynthesis, have brought about the current interest in neat polymer modification and functionalization *via* chemical, physical and enzymatic processes. The success of these processes largely depends on the extent to which the modification confers the desired trait(s) to the polymer allowing it to perform the targeted function efficiently. Depending on the type of PHA to be modified and its intended end uses, optimal modification calls for sensible manipulation of the process conditions such as catalyst loading, contact or exposure time, reactant or sample concentration, treatment dose *etc*. With rational and proper choice of modification strategies, a highly hydrophobic and crystalline PHA could be modified into a unique polymer with desired bioactive properties, a marked increase in wettability, elasticity and storage modulus characteristics, thereby allowing the functionalized polymers to be used in diverse novel applications.

References

1. R. Rai, D. M. Yunos, A. R. Boccaccini, J. C. Knowles, I. A. Barker, S. M. Howdle, G. D. Tredwell, T. Keshavarz and I. Roy, *Biomacromolecules*, 2011, **12**, 2126–2136.
2. A. L. Chung, H. L. Jin, L. J. Huang, H. M. Ye, J. C. Chen, Q. Wu and G. Q. Chen, *Biomacromolecules*, 2011, **12**, 3559–3566.
3. K.-S. Ng, Y.-M. Wong, T. Tsuge and K. Sudesh, *Process Biochem.*, 2011, **46**, 1572–1578.
4. A. M. Gumel, M. S. M. Annuar and T. Heidelberg, *PLoS One*, 2012, **7**, e45214.
5. A. M. Gumel, M. S. M. Annuar and T. Heidelberg, *Polym. Degrad. Stab.*, 2012, **97**, 1224–1231.

6. A. M. Gumel, M. S. M. Annuar and Y. Chisti, *Ultrason. Sonochem.*, 2013, **20**, 937–947.

7. A. M. Gumel, M. S. M. Annuar and Y. Chisti, *J. Polym. Environ.*, 2013, **21**, 580–605.

8. A. Poli, P. Di Donato, G. R. Abbamondi and B. Nicolaus, *Archaea*, 2011, **2011**, DOI: 10.1155/2011/693253.

9. A. M. Gumel, M. S. M. Annuar and T. Heidelberg, *J. Appl. Polym. Sci.*, 2013, **129**, 3079–3088.

10. S. Ray and M. Bousmina, *Prog. Mater. Sci.*, 2005, **50**, 962–1079.

11. C. Johansson, in *Nanocomposites with Biodegradable Polymers: Synthesis, Properties and Future Perspectives*, ed. V. Mittal, Oxford University Press, Oxford, 2011, pp. 348–367.

12. K. Fritzsche, R. W. Lenz and R. C. Fuller, *Int. J. Biol. Macromol.*, 1990, **12**, 92–101.

13. B. Hazer, R. Lenz and R. Fuller, *Macromolecules*, 1994, **27**, 45–49.

14. B. Hazer and A. Steinbüchel, *Appl. Microbiol. Biotechnol.*, 2007, **74**, 1–12.

15. M.-M. Bear, M.-A. Leboucher-Durand, V. Langlois, R. W. Lenz, S. Goodwin and P. Guérin, *React. Funct. Polym.*, 1997, **34**, 65–77.

16. T. Imamura, T. Kenmoku, T. Honma, S. Kobayashi and T. Yano, *Int. J. Biol. Macromol.*, 2001, **29**, 295–301.

17. T. Kenmoku, E. Sugawa, T. Yano, T. Nomoto, T. Imamura, T. Suzuki and T. Honma, Google Patents, 2004.

18. Y. B. Kim, R. Lenz and R. C. Fuller, *Macromolecules*, 1992, **25**, 1852–1857.

19. Y. Doi and C. Abe, *Macromolecules*, 1990, **23**, 3705–3707.

20. O. Kim, R. A. Gross, W. J. Hammar and R. A. Newmark, *Macromolecules*, 1996, **29**, 4572–4581.

21. C. Ewering, T. Lütke-Eversloh, H. Luftmann and A. Steinbüchel, *Microbiology*, 2002, **148**, 1397–1406.

22. *USA Pat.*, 5, 994, 478, 1999.

23. D. T. Shah, M. Tran, P. A. Berger, P. Aggarwal, J. Asrar, L. A. Madden and A. J. Anderson, *Macromolecules*, 2000, **33**, 2875–2880.

24. A. Althuri, J. Mathew, R. Sindhu, R. Banerjee, A. Pandey and P. Binod, *Bioresour. Technol.*, 2013, **145**, 290–296.

25. F. Kessler, L. Marconatto, R. da Silva Bussamara Rodrigues, G. A. Lando, A. Schank, M. H. Vainstein and D. E. Weibel, *J. Photochem. Photobiol. B: Biol.*, 2014, **130**, 57–67.

26. A. M. Gumel, M. S. M. Annuar and T. Heidelberg, *Int. J. Biol. Macromol.*, 2013, **55**, 127–136.

27. J. F. Ross, P. K. Chaudhuri and M. Ratnam, *Cancer*, 1994, **73**, 2432–2443.

28. A. M. Gumel, S. M. Annuar and T. Heidelberg, *J. Chem. Technol. Biotechnol.*, 2013, **88**, 1328–1335.

29. D. P. Martin, F. Skraly and S. F. Williams, Google Patents, 2003.

30. F. M. Samsuddin, R. L. Benjamin, Y. Sriram, A. Shant and W. S. Dennis, in *Green Polymer Chemistry: Biocatalysis and Materials II*, eds. H. N. H. N.

Cheng, R. A. Gross and P. B. Smith, American Chemical Society, 2013, vol. 1144, pp. 291–301.
31. D. Kai and X. J. Loh, *ACS Sustainable Chem. Eng.*, 2013, **2**(2), 106–119.
32. M. Y. Lee and W. H. Park, *Macromol. Chem. Phys.*, 2000, **201**, 2771–2774.
33. E. Renard, V. Langlois and P. Guérin, *Corros. Eng., Sci. Technol.*, 2007, **42**, 300–311.
34. N. Kurth, E. Renard, F. Brachet, D. Robic, P. Guerin and R. Bourbouze, *Polymer*, 2002, **43**, 1095–1101.
35. G. J. M. de Koning, H. M. M. van Bilsen, P. J. Lemstra, W. Hazenberg, B. Witholt, H. Preusting, J. G. van der Galiën, A. Schirmer and D. Jendrossek, *Polymer*, 1994, **35**, 2090–2097.
36. D. J. Stigers and G. N. Tew, *Biomacromolecules*, 2003, **4**, 193–195.
37. H. Arslan, B. Hazer and S. C. Yoon, *J. Appl. Polym. Sci.*, 2007, **103**, 81–89.
38. J. Babinot, E. Renard and V. Langlois, *Macromol. Chem. Phys.*, 2011, **212**, 278–285.
39. L. Zhou, Z. Chen, W. Chi, X. Yang, W. Wang and B. Zhang, *Biomaterials*, 2012, **33**, 2334–2344.
40. M. Kwiecień, G. Adamus and M. Kowalczuk, *Biomacromolecules*, 2013, **14**, 1181–1188.
41. D. Hu, A.-L. Chung, L.-P. Wu, X. Zhang, Q. Wu, J.-C. Chen and G.-Q. Chen, *Biomacromolecules*, 2011, **12**, 3166–3173.
42. A. P. Andrade, B. Witholt, R. Hany, T. Egli and Z. Li, *Macromolecules*, 2001, **35**, 684–689.
43. L. Timbart, E. Renard, M. Tessier and V. Langlois, *Biomacromolecules*, 2007, **8**, 1255–1265.
44. Z. Chen, S. Cheng, Z. Li, K. Xu and G.-Q. Chen, *J. Biomater. Sci., Polym. Ed.*, 2009, **20**, 1451–1471.
45. A. P. Andrade, P. Neuenschwander, R. Hany, T. Egli, B. Witholt and Z. Li, *Macromolecules*, 2002, **35**, 4946–4950.
46. Z. Chen, S. Cheng and K. Xu, *Biomaterials*, 2009, **30**, 2219–2230.
47. T. Imamura, T. Kenmoku, T. Honma, S. Kobayashi and T. Yano, *Int. J. Biol. Macromol.*, 2001, **29**, 295–301.
48. W. H. Park, R. W. Lenz and S. Goodwin, *J. Polym. Sci., Part A: Polym. Chem.*, 1998, **36**, 2381–2387.
49. W. H. Park, R. W. Lenz and S. Goodwin, *J. Polym. Sci., Part A: Polym. Chem.*, 1998, **36**, 2389–2396.
50. H. E. Kissinger, *Anal. Chem.*, 1957, **29**, 1702–1706.
51. T. Ozawa, *J. Therm. Anal. Calorim.*, 1970, **2**, 301–324.
52. W. H. Park, R. W. Lenz and S. Goodwin, *Polym. Degrad. Stab.*, 1999, **63**, 287–291.
53. R. D. Ashby, T. A. Foglia, D. K. Y. Solaiman, C.-K. Liu, A. Nuñez and G. Eggink, *Int. J. Biol. Macromol.*, 2000, **27**, 355–361.
54. M. Y. Lee, S. Y. Cha and W. H. Park, *Polymer*, 1999, **40**, 3787–3793.
55. M. Y. Lee and W. H. Park, *Polym. Degrad. Stab.*, 1999, **65**, 137–142.
56. A. H. Arkin, B. Hazer and M. Borcakli, *Macromolecules*, 2000, **33**, 3219–3223.

57. A. H. Arkin and B. Hazer, *Biomacromolecules*, 2002, **3**, 1327–1335.
58. C. Mihara, T. Yano, S. Kozaki, T. Honma, T. Kenmoku, T. Fukui and A. Kusakari, Google Patents, 2008.
59. T. Imamura, E. Sugawa, T. Yano, T. Nomoto, T. Suzuki, T. Honma, T. Kenmoku and T. Fukui, Google Patents, 2008.
60. H. W. K. Do Young Kim, M. G. Chung and Y. H. Rhee, *J. Microbiol.*, 2007, 87–97.
61. S. Nguyen, *Can. J. Chem.*, 2008, **86**, 570–578.
62. H.-K. Lao, E. Renard, I. Linossier, V. Langlois and K. Vallée-Rehel, *Biomacromolecules*, 2006, **8**, 416–423.
63. S. Ilter, B. Hazer, M. Borcakli and O. Atici, *Macromol. Chem. Phys.*, 2001, **202**, 2281–2286.
64. M. G. Chung, H. W. Kim, B. R. Kim, Y. B. Kim and Y. H. Rhee, *Int. J. Biol. Macromol.*, 2012, **50**, 310–316.
65. S.-G. Hu, C.-H. Jou and M.-C. Yang, *Carbohydr. Polym*, 2004, **58**, 173–179.
66. B. Hazer, *Int. J. Polym. Sci.*, 2010, **2010**, DOI: 10.1155/2010/423460.
67. M. Constantin, C. I. Simionescu, A. Carpov, E. Samain and H. Driguez, *Macromol. Rapid Commun.*, 1999, **20**, 91–94.
68. S. G. Hu, C. H. Jou and M. C. Yang, *J. Appl. Polym. Sci.*, 2003, **88**, 2797–2803.
69. W.-F. Lee and W.-J. Lin, *J. Polym. Res.*, 2002, **9**, 23–29.
70. J. Li, X. Li, X. Ni, X. Wang, H. Li and K. W. Leong, *Biomaterials*, 2006, **27**, 4132–4140.
71. N. A. Peppas, R. M. Ottenbrite, K. Park and T. Okano, *Biomedical applications of hydrogels handbook*, Springer, Germany, 2010.
72. B. Hazer, R. W. Lenz, B. Çakmaklı, M. Borcaklı and H. Koçer, *Macromol. Chem. Phys.*, 1999, **200**, 1903–1907.
73. K. Gagnon, R. Lenz, R. Farris and R. Fuller, *Polymer*, 1994, **35**, 4358–4367.
74. K. Gagnon, R. Lenz, R. Farris and R. Fuller, *Polymer*, 1994, **35**, 4368–4375.
75. A. S. Hoffman, *Chin. J. Polym. Sci.*, 1995, **13**, 195–195.
76. K. A. Thomas and S. D. Cook, *J. Biomed. Mater. Res.*, 1985, **19**, 875–901.
77. Y.-W. Wang, Q. Wu and G.-Q. Chen, *Biomaterials*, 2003, **24**, 4621–4629.
78. K. Burridge, K. Fath, T. Kelly, G. Nuckolls and C. Turner, *Annu. Rev. Cell Biol.*, 1988, **4**, 487–525.
79. C.-L. Dong, S.-Y. Li, Y. Wang, Y. Dong, J. Z. Tang, J.-C. Chen and G.-Q. Chen, *Biomaterials*, 2012, **33**, 2593–2599.
80. J. P. Vacanti and R. Langer, *The Lancet*, 1999, **354**, S32–S34.
81. X. Yang, K. Zhao and G.-Q. Chen, *Biomaterials*, 2002, **23**, 1391–1397.
82. R. D. Ashby, A.-M. Cromwick and T. A. Foglia, *Int. J. Biol. Macromol.*, 1998, **23**, 61–72.
83. H. W. Kim, C. W. Chung and Y. H. Rhee, *Int. J. Biol. Macromol.*, 2005, **35**, 47–53.
84. C. W. Chung, H. W. Kim, Y. B. Kim and Y. H. Rhee, *Int. J. Biol. Macromol.*, 2003, **32**, 17–22.

85. T. Hou, J. Zhang, L. Kong, X. Zhang, P. Hu, D. Zhang and N. Li, *J. Biomater. Sci., Polym. Ed.*, 2006, **17**, 735–746.
86. A. Tezcaner, K. Bugra and V. Hasırcı, *Biomaterials*, 2003, **24**, 4573–4583.
87. O. H. Kwon, I. S. Lee, Y.-G. Ko, W. Meng, K.-H. Jung, I.-K. Kang and Y. Ito, *Biomed. Mater.*, 2007, **2**, S52.
88. C. A. Woolnough, L. H. Yee, T. S. Charlton and L. J. R. Foster, *PLoS One*, 2013, **8**, e75817.
89. M. You, G. Peng, J. Li, P. Ma, Z. Wang, W. Shu, S. Peng and G.-Q. Chen, *Biomaterials*, 2011, **32**, 2305–2313.
90. Y. Dong, P. Li, C.-b. Chen, Z.-h. Wang, P. Ma and G.-Q. Chen, *Biomaterials*, 2010, **31**, 8921–8930.
91. M. Yamada, K. Yamashita, A. Wakuda, K. Ichimura, A. Maehara, M. Maeda and S. Taguchi, *J. Bacteriol.*, 2007, **189**, 1118–1127.
92. M. J. Humphries, S. K. Akiyama, A. Komoriya, K. Olden and K. M. Yamada, *J. Cell Biol.*, 1986, **103**, 2637–2647.
93. R. D. Ashby, T. A. Foglia, C.-K. Liu and J. W. Hampson, *Biotechnol. Lett.*, 1998, **20**, 1047–1052.
94. A. Dufresne, L. Reche, R. H. Marchessault and M. Lacroix, *Int. J. Biol. Macromol.*, 2001, **29**, 73–82.
95. F. Martellini, L. H. Innocentini Mei, S. Lora and M. Carenza, *Radiat. Phys. Chem.*, 2004, **71**, 257–262.
96. H. Mitomo, Y. Watanabe, I. Ishigaki and T. Saito, *Polym. Degrad. Stab*, 1994, **45**, 11–17.
97. M. González-Torres, A. M. Perez-González, M. González-Perez, C. Santiago-Tepantlán, S. G. Solís-Rosales and A. H. Heredia-Jiménez, *Int. J. Polym. Anal. Charact.*, 2011, **16**, 399–415.
98. M. G. Torres, J. C. Cortez, M. G. Pérez and R. R. Talavera, *Int. J. Sci. Adv. Technol.*, 2012, **2**(2), 106–119.
99. M. Bassas, A. M. Marqués and A. Manresa, *Biochem. Eng. J.*, 2008, **40**, 275–283.
100. M. Amiji and K. Park, *J. Biomater. Sci., Polym. Ed.*, 1993, **4**, 217–234.
101. Y.-Y. Shangguan, Y.-W. Wang, Q. Wu and G.-Q. Chen, *Biomaterials*, 2006, **27**, 2349–2357.
102. D. M. Zhang, F. Z. Cui, Z. S. Luo, Y. B. Lin, K. Zhao and G. Q. Chen, *Surf. Coat. Technol.*, 2000, **131**, 350–354.
103. N. J. Nosworthy, A. Kondyurin, M. M. M. Bilek and D. R. McKenzie, *Enzyme Microb. Technol.*, 2014, **54**, 20–24.
104. A. Belbah, A. Mkaddem, N. Ladaci, N. Mebarki and M. El Mansori, *Mater. Des.*, 2014, **53**, 202–208.
105. X.-Y. Chen, X.-F. Zhang, Y. Zhu, J. Zhang and P. Hu, *Polym. J.*, 2003, **35**, 148–154.
106. M. Manso, A. Valsesia, M. Lejeune, D. Gilliland, G. Ceccone and F. Rossi, *Acta Biomater.*, 2005, **1**, 431–440.
107. B. Santos, C. Rozsa, N. Galego, A. F. Michels, M. Behar and F. C. Zawislak, *Int. J. Polym. Anal. Charact.*, 2011, **16**, 431–441.

108. S. A. Mirmohammadi, M. T. Khorasani, H. Mirzadeh and S. Irani, *Polym.–Plast. Technol. Eng.*, 2012, **51**, 1319–1326.
109. N. De Geyter and R. Morent, in *Biomedical Science, Engineering and Technology*, ed. D. N. Ghista, 2012, Intech, Croatia, 225–246.
110. H. W. Kim, C. W. Chung, S. S. Kim, Y. B. Kim and Y. H. Rhee, *Int. J. Biol. Macromol.*, 2002, **30**, 129–135.
111. Y. Wang, L. Lu, Y. Zheng and X. Chen, *J. Biomed. Mater. Res., Part A*, 2006, **76**, 589–595.
112. V. Hasirci, A. Tezcaner, N. Hasirci and Ş. Süzer, *J. Appl. Polym. Sci.*, 2003, **87**, 1285–1289.
113. M. Nitschke, G. Schmack, A. Janke, F. Simon, D. Pleul and C. Werner, *J. Biomed. Mater. Res.*, 2002, **59**, 632–638.
114. A. Vallée, V. Humblot, R. A. Housseiny, S. Boujday and C.-M. Pradier, *Colloids Surf., B*, 2013, **109**, 136–142.
115. X. Li, K. L. Liu, M. Wang, S. Y. Wong, W. C. Tjiu, C. B. He, S. H. Goh and J. Li, *Acta Biomater.*, 2009, **5**, 2002–2012.
116. J. Zhan, X. Tian, Y. Zhu, L. Wang and L. Ren, *Appl. Surf. Sci.*, 2013, **280**, 564–571.
117. H. Yang, M. Sun, P. Zhou, L. Pan and C. Wu, *J. Biomed. Sci. Eng.*, 2010, **3**, 1146–1155.
118. X.-H. Qu, Q. Wu, J. Liang, X. Qu, S.-G. Wang and G.-Q. Chen, *Biomaterials*, 2005, **26**, 6991–7001.
119. Z. Karahaliloğlu, M. Demirbilek, M. Şam, M. Erol-Demirbilek, N. Sağlam and E. B. Denkbaş, *J. Appl. Polym. Sci.*, 2012, **128**, 1904–1912.
120. J. Zhang, K. Kasuya, A. Takemura, A. Isogai and T. Iwata, *Polym. Degrad. Stab.*, 2013, **98**, 1458–1464.
121. W. Yu, C.-H. Lan, S.-J. Wang, P.-F. Fang and Y.-M. Sun, *Polymer*, 2010, **51**, 2403–2409.
122. H. Fong, I. Chun and D. H. Reneker, *Polymer*, 1999, **40**, 4585–4592.
123. I. S. Lee, O. H. Kwon, W. Meng, I.-K. Kang and Y. Ito, *Macromol. Res.*, 2004, **12**, 374–378.
124. T. H. Ying, D. Ishii, A. Mahara, S. Murakami, T. Yamaoka, K. Sudesh, R. Samian, M. Fujita, M. Maeda and T. Iwata, *Biomaterials*, 2008, **29**, 1307–1317.
125. A. J. Anderson and E. A. Dawes, *Microbiol. Rev.*, 1990, **54**, 450–472.
126. H.-M. Müller and D. Seebach, *Angew. Chem., Int. Ed. Engl.*, 1993, **32**, 477–502.
127. S. Y. Lee, Y. Lee and F. L. Wang, *Biotechnol. Bioeng.*, 1999, **65**, 363–368.
128. Q. Ren, A. Grubelnik, M. Hoerler, K. Ruth, R. Hartmann, H. Felber and M. Zinn, *Biomacromolecules*, 2005, **6**, 2290–2298.
129. D. Jendrossek and R. Handrick, *Annu. Rev. Microbiol.*, 2002, **56**, 403–432.
130. J. Gangoiti, M. Santos, M. A. Prieto, I. de la Mata, J. L. Serra and M. J. Llama, *Appl. Environ. Microbiol.*, 2012, **78**, 7229–7237.

131. Y. Shirakura, T. Fukui, T. Saito, Y. Okamoto, T. Narikawa, K. Koide, K. Tomita, T. Takemasa and S. Masamune, *Biochim. Biophys. Acta, Gen. Subj.*, 1986, **880**, 46–53.

132. A. Schirmer, D. Jendrossek and H. G. Schlegel, *Appl. Environ. Microbiol.*, 1993, **59**, 1220–1227.

133. A. Schirmer and D. Jendrossek, *J. Bacteriol.*, 1994, **176**, 7065–7073.

134. M. Knoll, T. M. Hamm, F. Wagner, V. Martinez and J. Pleiss, *BMC Bioinf.*, 2009, **10**, 89.

135. K. E. Jaeger, A. Steinbuchel and D. Jendrossek, *Appl. Environ. Microbiol.*, 1995, **61**, 3113–3118.

136. K. Mukai, Y. Doi, Y. Sema and K. Tomita, *Biotechnol. Lett.*, 1993, **15**, 601–604.

137. Y. Tokiwa and T. Suzuki, *Nature*, 1977, **270**, 76–78.

138. I. H. Segel, *Biochemical calculations: how to solve mathematical problems in general biochemistry*, Wiley, New Jersey, 1976.

139. J. Gangoiti, M. Santos, M. J. Llama and J. L. Serra, *Appl. Environ. Microbiol.*, 2010, **76**, 3554–3560.

140. T. Kawata and H. Ogino, *Biotechnol. Prog.*, 2009, **25**, 1605–1611.

141. M. Z. Kamal, P. Yedavalli, M. V. Deshmukh and N. M. Rao, *Protein Sci.*, 2013, **22**, 904–915.

142. E. Antonian, *Lipids*, 1988, **23**, 1101–1106.

143. E. Santaniello, P. Ferraboschi and P. Grisenti, *Enzyme Microb. Technol.*, 1993, **15**, 367–382.

144. G. D. Yadav and K. M. Devi, *Chem. Eng. Sci.*, 2004, **59**, 373–383.

145. Y. Ikada and H. Tsuji, *Macromol. Rapid Commun.*, 2000, **21**, 117–132.

146. D.-E. Ch'ng and K. Sudesh, *AMB Express*, 2013, **3**, 1–11.

147. S. H. Ha, N. M. Hiep, S. H. Lee and Y. M. Koo, *Bioprocess Biosyst. Eng.*, 2010, **33**, 63–70.

148. S. Mat Radzi, M. Basri, A. Bakar Salleh, A. Ariff, R. Mohammad, M. B. Abdul Rahman and R. N. Z. R. Abdul Rahman, *Electron. J. Biotechnol.*, 2005, **8**, 291–298.

149. M. Kitagawa and Y. Tokiwa, *Biotechnol. Lett.*, 1998, **20**, 627–630.

150. L. Pastorino, F. Pioli, M. Zilli, A. Converti and C. Nicolini, *Enzyme Microb. Technol.*, 2004, **35**, 321–326.

151. G. Sekosan, *Effect of Microstructure of Poly-e-caprolactone on Enzymatic Degradation*, MSc Thesis, Long Island University, The Brooklyn Center, 2009.

152. G. Q. Chen and Q. Wu, *Appl. Microbiol. Biotechnol.*, 2005, **67**, 592–599.

153. J. F. Kennedy and Z. M. B. Figueiredo, *Carbohydr. Polym.*, 1994, **23**, 76.

154. S. Sato, Y. Ono, Y. Mochiyama, E. Sivaniah, Y. Kikkawa, K. Sudesh, T. Hiraishi, Y. Doi, H. Abe and T. Tsuge, *Biomacromolecules*, 2008, **9**, 2811–2818.

155. O. Tasaki, A. Hiraide, T. Shiozaki, H. Yamamura, N. Ninomiya and H. Sugimoto, *JPEN, J. Parenter. Enteral Nutr.*, 1999, **23**, 321–325.

156. S. Nguyen and R. H. Marchessault, *Macromol. Biosci.*, 2004, **4**, 262–268.

157. T. Shinoka, D. Shum-Tim, P. X. Ma, R. E. Tanel, N. Isogai, R. Langer, J. P. Vacanti and J. E. Mayer, *J. Thorac. Cardiovasc. Surg.*, 1998, **115**, 536–545.

158. C. Chen, B. Fei, S. Peng, H. Wu, Y. Zhuang, X. Chen, L. Dong and Z. Feng, *J. Polym. Sci., Part B: Polym. Phys.*, 2002, **40**, 1893–1903.

159. Y. J. Kim, I.-K. Kang, M. W. Huh and S.-C. Yoon, *Biomaterials*, 2000, **21**, 121–130.

160. Y. Ito, M. Inoue, S. Q. Liu and Y. Imanishi, *J. Biomed. Mater. Res.*, 1993, **27**, 901–907.

161. S. Q. Liu, Y. Ito and Y. Imanishi, *J. Biomed. Mater. Res.*, 1993, **27**, 909–915.

162. I. K. Kang, S. H. Choi, D. S. Shin and S. C. Yoon, *Int. J. Biol. Macromol.*, 2001, **28**, 205–212.

163. J. Ihssen, D. Magnani, L. Thöny-Meyer and Q. Ren, *Biomacromolecules*, 2009, **10**, 1854–1864.

164. A. Shrivastav, H. Y. Kim and Y. R. Kim, *BioMed Res. Int*, 2013, **2013**, 581684.

165. H.-N. Kim, J. Lee, H.-Y. Kim and Y.-R. Kim, *Chem. Commun.*, 2009, 7104–7106.

166. H.-j. Paik, Y.-R. Kim, R. N. Orth, C. K. Ober, G. W. Coates and C. A. Batt, *Chem. Commun.*, 2005, 1956–1958.

167. B. H. A. Rehm, R. V. Antonio, P. Spiekermann, A. A. Amara and A. Steinbüchel, *Biochim. Biophys. Acta, Protein Struct. Mol. Enzymol.*, 2002, **1594**, 178–190.

168. J. Lee, S.-G. Jung, C.-S. Park, H.-Y. Kim, C. A. Batt and Y.-R. Kim, *Bioorg. Med. Chem. Lett.*, 2011, **21**, 2941–2944.

169. H. P. Molitoris, S. T. Moss, G. J. M. de Koning and D. Jendrossek, *Appl. Microbiol. Biotechnol.*, 1996, **46**, 570–579.

170. Y.-W. Wang, W. Mo, H. Yao, Q. Wu, J. Chen and G.-Q. Chen, *Polym. Degrad. Stab.*, 2004, **85**, 815–821.

171. A. M. Gumel, M. S. M. Annuar, Y. Chisti and T. Heidelberg, *Ultrason. Sonochem.*, 2012, **19**, 659–667.

172. Z. Gan, Q. Liang, J. Zhang and X. Jing, *Polym. Degrad. Stab.*, 1997, **56**, 209–213.

173. N. Ansari and A. A. Amirul, *Appl. Biochem. Biotechnol.*, 2013, **170**, 690–709.

174. Y. Kanesawa, N. Tanahashi, Y. Doi and T. Saito, *Polym. Degrad. Stab.*, 1994, **45**, 179–185.

175. S. F. Williams, D. P. Martin, D. M. Horowitz and O. P. Peoples, *Int. J. Biol. Macromol.*, 1999, **25**, 111–121.

176. S. F. Williams, D. P. Martin, D. M. Horowitz and O. P. Peoples, *Int. J. Biol. Macromol.*, 1999, **25**, 111–121.

177. M. K. Alexander, *Nature*, 2001, **409**, 241–246.

178. A. M. Gumel, M. S. M. Annuar, T. Heidelberg and Y. Chisti, *Process Biochem.*, 2011, **46**, 2079–2090.

179. A. M. Gumel, M. S. M. Annuar, T. Heidelberg and Y. Chisti, *Bioresour. Technol.*, 2011, **102**, 8727–8732.

180. A. Zaks and A. M. Klibanov, *Proc. Natl. Acad. Sci. U. S. A.*, 1985, **82**, 3192–3196.
181. A. Córdova, *Biomacromolecules*, 2001, **2**, 1347–1351.
182. A. M. Gumel, S. M. Annuar and T. Heidelberg, *J. Chem. Technol. Biotechnol.*, 2013, **88**, 1328–1335.
183. F. Ravenelle and R. H. Marchessault, *Biomacromolecules*, 2002, **3**, 1057–1064.
184. S. Dai and Z. Li, *Biomacromolecules*, 2008, **9**, 1883–1893.
185. C. Laane, S. Boeren, K. Vos and C. Veeger, *Biotechnol. Bioeng.*, 1987, **30**, 81–87.
186. M. Kamachi, S. Zhang, S. Goodwin and R. W. Lenz, *Macromolecules*, 2001, **34**, 6889–6894.
187. B. Kunasundari and K. Sudesh, *Express Polym. Lett.*, 2011, **5**, 620–634.
188. H. Ogino and H. Ishikawa, *J. Biosci. Bioeng.*, 2001, **91**, 109–116.
189. M. P. Stevens, *Polymer Chemistry: An Introduction*, Oxford University Press, USA, 3rd edn, 2009.
190. A. Tanaka and T. Kawamoto, *Bioprocess Technol.*, 1991, **14**, 183–208.

CHAPTER 8

Polyhydroxyalkanoates as Packaging Materials: Current Applications and Future Prospects

LACHLAN HARTLEY YEE[a] AND LESLIE JOHN RAY FOSTER*[b]

[a] Marine Ecology Research Centre, School of Environment, Science and Engineering, Southern Cross University, Lismore NSW 2480, Australia; [b] Bio/Polymer Research Group, School of Biotechnology & Biomolecular Sciences, University of New South Wales, Sydney NSW 2052, Australia
*Email: J.Foster@unsw.edu.au

8.1   Introduction

Packaging is generally defined in government regulations as:

All products made of any materials of any nature to be used for the containment, protection, handling, delivery and preservation of goods from the producer to the user or consumer.[1]

Packaging can be classified from primary to tertiary dependent upon function and size (Table 8.1). Primary packaging materials for items at the point of sale traditionally consist of plastics, cardboard, aluminium and tin, while consolidation of primary items are usually in the form of cardboard or wooden boxes, with and without various protective supports, such as

RSC Green Chemistry No. 30
Polyhydroxyalkanoate (PHA) Based Blends, Composites and Nanocomposites
Edited by Ipsita Roy and Visakh P M
© The Royal Society of Chemistry 2015
Published by the Royal Society of Chemistry, www.rsc.org

Table 8.1   Table of different packaging classifications with examples.

I°	Sales packaging		Packaging that forms a sales unit for the final consumer or user, so-called 'sales' packaging.	A plastic yoghurt pot.
II°	Grouped packaging		Packaging that contains a number of primary sales units.	Cardboard box containing plastic yoghurt pots.
III°	Transport packaging		Tertiary packaging that is used to group secondary packaging together to aid handling and transportation and to prevent product damage.	Wooden pallet and plastic shrink-wrap containing boxes of yoghurt pots.

air-filled foams and 'polystyrene beads'. Large volume transports mainly rely upon wood or plastic pallets with plastic shrink-wrap to consolidate multiple items.

The growing use of e-commerce and on-line auctions has made the business of product sourcing much simpler, promoting cost efficiency for users of consumables, while driving an increase in the amount of packaging being used. The global packaging market increased steadily from US$ 372 billion in 1999 to US$ 563 billion in 2009.[2]

Plastics are synthetic organic polymers derived from the polymerisation of monomers extracted from fossil fuels, oil or gas.[3,4] Since the initial development of 'Bakelite' in 1907, the first modern plastic, a variety of inexpensive manufacturing techniques have been optimised and have led to the mass production of a number of plastics exhibiting lightweight, inert and corrosion-resistant properties.[5] These attributes have resulted in the extensive use of plastics in a plethora of inexhaustible applications.[5] Since the beginning of mass production in the 1940s, the amount of plastic being manufactured has increased rapidly. In 2009, 230 million tonnes of plastic were produced globally and accounted for approximately 8% of global oil production.[4,6] In 2003, the world market for plastic packaging was valued at US$ 130 billion and has been growing at an annual rate of 5% since then.[7]

Originally developed to be more durable than natural products, plastics were never designed for today's disposable culture. It is testament to their adaptability that plastics currently play such an important role in packaging. The Environmental Protection Agency (EPA) of the USA reports that in 2011, 14 million tonnes of plastic packaging constituted 44% of plastic waste and 5.6% of all municipal solid waste (MSW).[2] In the same year, the European Union reports that plastics constituted 19% by weight (14.9 million tonnes) of all packaging waste, with each citizen in the 27 member states producing 159 Kg of packaging waste in the year.[8]

While some plastic waste is recycled, the majority is disposed of through landfilling, where it can take hundreds of years to break down, or through incineration, which can release toxic fumes.[9,10] Furthermore, there is growing concern with the environmental impact of persistent plastic packaging that fails to enter any waste streams, for example 'The Great Pacific Garbage Patch', estimated to contain more than 3 million tonnes of plastic.[11] In 2006, Thomson estimated that 10% by weight of all plastics produced ended up in the world's oceans.[12]

It is becoming increasingly clear that there is a definite need for environmentally-friendly, sustainable packaging materials to protect environmental and public health. Despite this, the EU reports that the use of plastic had the highest growth rate of all packaging materials, increasing from 17.9 to 18.7% from 2005 to 2011.[8] In this chapter, we review the role polyhydroxyalkanoates (PHAs) currently play in packaging as well as summarise the research being undertaken to increase their usage in this application field.

8.2 The Need for Sustainable Packaging Materials

Arguably, the two largest driving forces behind the need to adopt biopolymer sources for sustainable packaging are: (a) the depleting reserve of fossil fuels as a source of synthetic polymers for packaging and (b) the pollution caused by non-degradable polymers in addition to (c) the carbon footprint of waste fossil fuel derived polymers that are eventually incinerated. Indeed, the formation of packaging from fossil fuel sources, referred to here as ff-polymers, is merely a storage bank for fossil fuels themselves on their way to incineration for energy.

Recycling of plastic packaging has offered a potential route to impose some sustainability on the use of conventional plastics. However, recycling rates can vary widely from country to country. For 2011, the EU reports a recycling rate of 22.5%, while in the USA, the EPA reports an overall plastics recycling rate of 8% (2.7 million tonnes).[2,8] In contrast, Australia has a strong dependence upon landfilling for its waste management, with nearly 60% of all MSW generated in 2006–7 being disposed of in this manner.[13] Nevertheless, in 2003, the overall plastic packaging recycling rate in Australia was 20.5%.[14] Despite significant efforts being made to legislate and increase plastic recycling, the World Packaging Organisation reports that 36% of industry professionals regarded 'recycling as being of no importance to the market'.[7]

Recycling rates are also greatly influenced by the type of plastic and its application, thus in 2011, 29% of high density poly(ethylene), HDPE, bottles and 29% of poly(ethylene terephthalate), PET, bottles and jars in the USA were recycled. The primary market for plastics from recycled PET bottles is as fibres for carpet and textiles, while that for recycled HDPE is bottles.[2]

While producing plastics from recycled materials is estimated to use only 30% of the energy required to make plastic products from fossil fuels, the advantage of biopolymer derived packaging is a net zero production of additional atmospheric CO_2 as well as other associated greenhouse gases, such as SO_x, NO_x and methane. In response to the negative environmental impact of current packaging materials, comparatively stricter governmental regulations are providing a primary driver for sustainable and biodegradable packaging.[15]

A holistic design of packaging is required to rethink the approach in which sustainable resources through biopolymer-led packaging will be incorporated into existing packaging needs. Colwill and co-authors have combined the science of business planning and the chemistry and engineering design of packaging to propose a holistic or ground up design to support effective development of biopolymer centred packaging.[16] Their model does not purely focus on economic gain but also prioritises a long term environmental and sustainable model that is a desirable goal if we are to embrace technology that releases us from fossil fuel dependence.

8.2.1 Commercial Plastic Additives

While current plastics based on ff-polymers are generally considered biochemically inert, they can contain a variety of additives.[17] Additives in the form of plasticising agents, flame retardants and colourants are incorporated into plastics during manufacture to modify their appearance as well as their mechanical and thermal properties. For example, polybrominated diphenyl ethers are usually added to extend the life of plastics by providing heat resistance; nonylphenol can be added to prevent oxidative damage while triclosan has actually been used to prevent biodegradation.[18] Plastic additives are of considerable environmental concern as they leach out into the environment, introducing hazardous chemicals to biota.[19]

The rate at which these additives leach from plastics is dependent on the pore size of the polymer matrix. Pore sizes vary according to the polymer, the properties and size of the additive and environmental conditions.[10,20] Phthalates are emollients, additives used to soften plastics by reducing the affinity between the molecular chains within the synthetic polymer matrix, and can constitute up to 50% by weight in poly(vinyl chloride), (PVC).[21] Similarly, bisphenol A (BPA) is a polycarbonate additive that has been widely used in food and beverage containers. Neither of these additives are persistent, however, their instability within plastic products facilitates their leaching with a loss of material performance. Furthermore, vom Saal and Myers have reported the high prevalence of both BPA and phthalates in aquatic environments, particularly in landfill leachates.[22]

In 1996, Colborn *et al.* highlighted the potential health implications of these additives and their ability to disrupt the natural function of the endocrine system.[23] The number of suspected endocrine disrupting chemicals (EDCs) has subsequently grown.[24] However, the causality between exposure to EDCs and adverse human health effects is unclear and controversial due to the multifactorial etiologies of such a hormone-related disease.[25,26]

Swan *et al.* report that *in utero* exposure to environmental levels of phthalate plasticisers is associated with a decreased anogenital distance in male infants, indicating a significant reduction in virilisation.[27] Similarly, Howdeshell *et al.* reported that exposure to BPA induced an earlier onset of puberty in female mice, while Colón *et al.* have suggested that phthalate exposure in female infants also correlates with an earlier onset of puberty.[28,29] To complicate matters and fuel the debate, a number of EDCs have been found to possibly exhibit epigenetic transgenerational effects.[30]

8.3 Biodegradable Plastics

The International Standards Organisation (ISO) is continuously developing and revising material standards and their tests, but at the time of writing, the ISO lists over 160 existing standards relating to plastics, with five regarding their biodegradability, and a further two are under development (Table 8.2). According to the ISO, biodegradable plastics are defined as those that undergo significant changes in chemical structure under specific environmental conditions. Biodegradation involves microbial enzymatic action

Table 8.2 Table of existing and developing ISO standards regarding biodegradability of plastics.

ISO Standard	Title	Date
14853:2005	Plastics - Determination of the ultimate anaerobic biodegradation of plastic materials in an aqueous system - Method by measurement of biogas production	2nd February 2005
14855-2:2007	Determination of the ultimate aerobic biodegradability of plastic materials under controlled composting conditions - Method by analysis of evolved carbon dioxide - Part 2: Gravimetric measurement of carbon dioxide evolved in a laboratory-scale test	28th August 2007
13975:2012	Plastics - Determination of the ultimate anaerobic biodegradation of plastic materials in controlled slurry digestion systems - Method by measurement of biogas production	4th May 2012
10210:2012	Plastics - Methods for the preparation of samples for biodegradation testing of plastic materials	7th August 2012
17566:2012	Plastics - Determination of the ultimate aerobic biodegradability of plastic materials in soil by measuring the oxygen demand in a respirometer or the amount of carbon dioxide evolved	8th August 2012
ISO/CD 18830	Plastics - Test method for determining aerobic biodegradation of plastic materials sunk at the sea water/sandy sediment interface	*Under Development*
ISO/NP 19679	Plastics - Determination of aerobic biodegradation of non-floating plastic materials in a seawater/sediment interface - Method by analysis of evolved carbon dioxide	*Under Development*

as opposed to abiotic weathering. In order to be considered and marketed as biodegradable, plastics must satisfy ISO standards as per Table 8.2.[31]

8.4 Polyhydroxyalkanoates (PHAs)

Polyhydroxyalkanoates (PHAs) are a family of biopolyesters synthesised by a diverse range of microorganisms under conditions of excess carbon and one or more limiting nutrients.[32] Under these unbalanced conditions, PHAs are stored as discrete intracellular inclusion bodies where they serve as an energy storage system in a similar manner to glycogen in mammals; in certain genera they may also serve as an ion sink and play an important role in encystment and sporulation.[33,34] Microorganisms subsequently catabolise PHAs into carbon and hydrogen when needed. Flexibility in the bioprocessing systems, in particular the carbon source, leads to different monomeric compositions; at the last count, there were over approximately 150 different monomeric components.[35] By far the greatest group of PHAs occur with variations within the side chain of the monomer (Figure 8.1(a)). Although, more recently there have been a number of PHAs possessing variations in the back bone length.[35]

   PHAs are classified according to the number of monomeric carbons, those of relatively short chain length (sclPHAs) with three to six carbons, those with seven to 16 carbons as mclPHAs (medium chain length) and PHAs with monomers possessing more than 16 carbons as long chain length, lclPHAs (Figure 8.1). In addition to simple variations in the length of the chain backbone and side chain, a number of PHAs containing functional chemical groups have also been synthesised. These include halogens, branched and aromatic as well as unsaturated groups.[35] The most common PHA packaging resins are polyhydroxybutyrate (PHB) and its copolymer with poly-hydroxyvalerate (PHB-*co*-HV), (Figure 8.1(b), (c)).

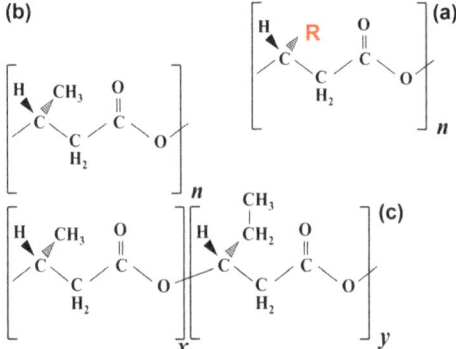

Figure 8.1   Chemical formulae for the family of polyhydroxyalkanoates (a) and the two most common commercial plastic resins, polyhydroxybutyrate (b) and its copolymer with polyhydroxyvalerate (c).

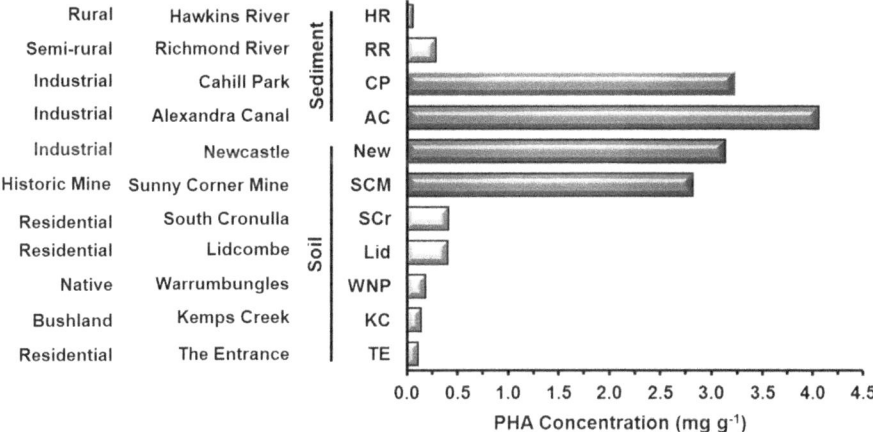

Figure 8.2 Graph showing the change in environmental concentrations of PHAs from different environments sampled in New South Wales, Australia.

The potential of PHAs as biodegradable replacements for conventional bulk commodity plastic packaging while promoting sustainable development has long been recognised.[36] However, despite a body of research reporting their environmental degradation, there is comparatively little data on their environmental occurrence.[36–40] While the biota from and in environmental samples can be induced to synthesise significant concentrations of PHAs—up to 90% of their cellular dry weight—'natural' concentrations of PHAs are significantly lower.[37–41] In their research measuring environmental PHA concentrations in sediment, Findlay and White (1983) speculated that PHAs might be used to elucidate the nutrient history of the sampling environment.[42] Similarly, Foster *et al.* measured PHA concentrations in soils and sediment samples from pristine environments and those known to have undergone recent pollution (Figure 8.2).[43] Their research showed background levels of PHAs of 0.12 to 0.40 mg g^{-1} in soil from unpolluted environments characterised by agricultural land or native vegetation. In contrast, PHAs in samples affected by anthropogenic activity displayed concentrations 14 to 40 times higher. The ubiquitous presence of PHAs, as determined in a variety of environments, is supportive of their development as bioplastics for environmentally-friendly, biodegradable packaging.[43]

8.4.1 Commercial PHA Packaging

The potential of PHAs for truly biodegradable packaging was recognised in the 1980s with the commercial release of Biopol®, thermoplastic resins of P(3HB) with various copolymer loadings of (3HV), by Imperial Chemical Industries (ICI, now Zeneca). The biodegradability of PHAs stems from the fact that PHAs will break down 'fully' to water and carbon dioxide. The environmentally-friendly impact of PHAs also stems from their microbially derived source and they are therefore considered a biopolymer among the categories of biopolymer production.[44] These can be summarised as follows:

(a) Polymers from biomass such as the agro-polymers from agro-resources, *e.g.*, starch, cellulose.
(b) Polymers obtained by microbial production, *e.g.*, poly(hydroxyalkanoates).
(c) Polymers chemically synthesized using monomers obtained from agro-resources, *e.g.*, poly(lactic acid).
(d) Polymers whose monomers and polymers are both obtained by chemical synthesis from fossil resources, *e.g.*, poly(caprolactone), polyester amide, *etc.*

The first commercialisation of Biopol® was a trial release in 1992 of the 'world's first totally biodegradable product', a Wella™ shampoo in an extrusion moulded bottle (Figure 8.3). However, despite a public willingness to embrace this environmentally-friendly product, high production costs made long term sustainability uneconomic and rights were eventually transferred to Metabolix in 2001 (through Monsanto in 1996).

Currently, there are a number of commercial suppliers of PHAs for application as packaging materials (Table 8.3). All the companies use vegetable-based feedstocks with sugar beet predominating. Biomer, Goodfellow and Procter & Gamble Chemicals (P&G) produce P(3HB), while all companies with the exception of Goodfellow, also produce PHA copolymers with various loadings of 3-hydroxyvalerate (3HV), 3-hydroxyhexanoate (3HHx) or 4-hydroxybutyrate (4HB). More recently, Proganic has manipulated thermoplastic properties and associated economics by blending PHAs with PLA for its resin, Proganic®, while NaturePlast also have PHA-based resins ranging from 100 to 55% 'bio-based' (PHI 001).

Manipulation of physical and material properties through PHA composition and molecular properties, as well as blending, is standard practice. Metabolix's Mirel™ P3001/F3002 has suitable mechanical flexibility, a relatively high melt strength and is acceptable for food contact (Figure 8.4).

Figure 8.3   Wella shampoo bottle produced from biodegradable PHA resins of PHB and P(HB-*co*-HV), marketed by ICI as Biopol™, 'Natures Plastic' (image supplied by Foster).

Table 8.3 Current commercial suppliers of PHA-based thermoplastic resins used for packaging, and their product range.

Company	Product Range
Biomer	P209, P300, P226, P304
Goodfellow	Polyhydroxybutyrate
Metabolix	M-Vera™ B5002, B5010
	Mirel™ P5001, P1003/F1005, P4010
NaturePlast	PHC001, PHE003, PHE008, PHE001, PHI008
Bi-On	Minerv-SB™, -SC™
P&G Chemicals/Meridian Inc.	Nodax™
Proganic	Proganic®
TianAn Biologic Materials Co. Ltd.	ENMAT™ Y1000P

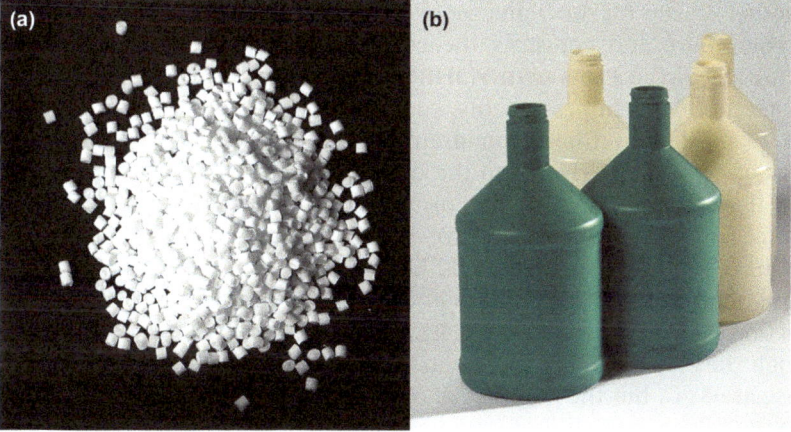

Figure 8.4 Photographs of PHA resin supplied as pellets (a) and as blow moulded bottles (b) (images supplied by Mirel™).

Its subsequent FDA approval has led to its use in a wide variety of applications for food packaging, including compost bags, yoghurt containers, tubs and food trays as well as sealable cups and pots. Its thermal properties allow packaged goods to be frozen and microwaved up to 100 °C, while its biodegradability supports sustainability. Mirel™'s thermal properties support its processing by extrusion moulding, while a range of NaturePlast resins, also used in food packaging, have properties more suitable to injection moulding. In contrast, NaturePlast PHE 008, while advertised as an 80% bio-based PHA resin, is not biodegradable but compostable and can also be processed through vacuum foaming. Nodax™, P(3HB-*co*-3HHx), is currently used for detergent bottles, coffee containers and foam containers.

8.5 Factors Affecting PHA Production

One of the prohibiting factors against the use of PHAs as a resin for packaging materials is that it is economically uncompetitive in the current market compared to fossil fuel sourced synthetic raw material (ff-polymers).[45,46] In the production of PHA, the cost of the carbon substrate represents approximately 50% of the total production cost.[45] Thus, one strategy to reduce this substantial expense would be to source carbon feed-stocks from relatively cheaper sources. Examples of relatively cheaper carbon sources have been reviewed by Sreekanth and co-workers and include date syrup, starch, and mahuva flower extract as examples.[45] Since PHAs are microbially derived materials, a more economical food source for production bacteria would contribute to a lowering of the PHA production cost.[45] Sreekanth *et al.* exploited a *Bacillus* sp., CFR 67, for not only the production of PHA but simultaneous production of alpha-amylase. They found that wheat bran hydrolysate from agricultural waste products was the most productive substrate with 73 U mL^{-1} amylase and 524 mg L^{-1} PHA being produced.[45]

Another reason for the high cost of PHA production was described by Chen and co-authors and concerns the costs associated with the recovery and purification of the PHAs with P(3HB), for example, being as high as 50%.[46] Thus methodology improvements, such as the aqueous extraction method of P(3HB) reported by Chen, can dramatically reduce costs.[46] More efficient means of extracting PHAs from the biological hosts would enhance the economical attraction of these biopolymers for investors and manufacturers. More recently, Chen and Zhang have investigated how a moderately halophilic bacterium, *Halomonas salina,* could be employed for P(3HB) production with the use of an osmotic down-shock aqueous method selected for cost-effective extraction.[46] A very encouraging 92.3% extraction efficiency was achieved through the combined use of osmotic down-shock with additional ultrasonic release at 65 °C, but the addition of NaCl (6%) and SDS (0.5%) was required, while the solvent was chloroform. A comprehensive review of recent advances in the production of PHAs has been published by Gumel and co-authors.[47]

Cost-effective production of PHAs using transgenic systems continues to be an attractive option. In 1994, Somerville and co-workers successfully transferred the genes for PHA synthesis from the bacterium *Ralstonia eutropha* into the plant *Aradopsis thaliana*, targeting their expression in plant leaves where PHA production attained 14% of material dry weight.[48] However, while PHA production using photosynthetic systems would benefit economically from existing crop production and harvesting technologies, it has failed to win public opinion in the ongoing GMO debate.

8.6   Properties of PHAs for Packaging

In order to rival current synthetic polymers used in packaging, such as poly-ethylene (PE), polypropylene (PP) or polystyrene (PS) for example, PHAs' suite of both physical and chemical properties need to be comparable (Figure 8.5). Here, we describe the various mechanical, thermal and optical properties afforded by PHA and how they compare to their synthetic counterparts. Table 8.4 provides an example of both tensile and thermal properties for a variety of commercially available PHAs.

Table 8.4 Examples of material properties for some commercial PHA-based resins used for packaging, illustrating the variation in mechanical, thermal and processing properties.

	Tensile elongation (%)	Tensile modulus (MPa)	Tensile strength (MPa)	Melt mass flow rate (g/10 min)	Melting point (°C)	Heat deflection temperature (°C)
GoodFellow						
PHB	—	3.5 (GPA)	40		—	—
NaturePlast						
PHI004	20	1650	—	4	—	—
PHI003	2	2950	—	15–30	—	—
PHI005	300	2200	—	15–20	—	—
PHC001	—	—	—	200 000	—	—
PHI001 (55% Bio-based)	3.5	860	14.2	10–15	145–155	45
PHE003 (80% ")	400	800	—	3	—	—
PHE008 (80% ")	13	—	19	3	100–190	116
PHI008 (90% ")	4	2780	25	30	100–190	118
Biomer						
P300	11	1850	28	—	—	108
P304	8	1500	28	—	—	108
P209	11–18	900–1200	15–20	10 KJ m^{-2}	—	—
P226	6–9	1700–2000	24–27	10 KJ m^{-2}	—	—
Metabolix						
Mirel P4010	10	—	10	—	—	—
Mirel P3001/ F3002	13	—	19	—	—	116

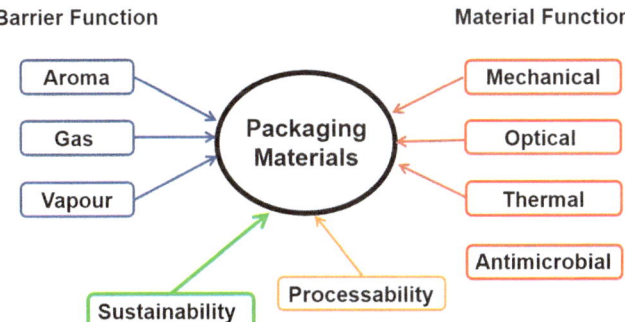

Figure 8.5 Diagrammatic flow chart of factors influencing the design of packaging materials.

8.6.1   Mechanical Properties

Compared to conventional, synthetically derived plastics, the mechanical properties of PHA biopolymers are often seen as inferior.[49] The argument for composite materials, and indeed where the additive is of a nano-dimension, is proposed as a solution that produces mechanical properties far more desirable from a packaging point of view.[49] The nano-dimension additives have the extra advantage of possessing a high surface area and therefore increasing the matrix and additive interaction. However, the ultimate fate of these nano-particles has raised concerns and is an area worthy of greater study.

Further support for nanocomposite blends involving PHA was highlighted by the study of Ohashi *et al.* in which they aimed to produce packaging films by blending P(3HB), polyethylene (PE) and colloidal silica. Blending was achieved in a single-screw extruder and the mechanical properties proved to be of great interest. For a small amount of added silica (0.2 to 0.4% where moles or wt% is not specified by the article), the tensile strength increased significantly, by 20% in magnitude.[50] Importantly, the improvement in mechanical properties thanks to nano-composite blending highlights that properties desired within PHB can often be tuned as needed with additive polymers such as PE and SiO_2 but ideally the additives should be from renewable sources in their own right.

8.6.2   Optical Properties

The optical properties of plastic packaging, certainly in the case of food packaging, offer a convenient, lightweight and flexible adaption of packaging technologies for the food industry, reducing the reliance upon glass and metallic canning. Transparency, a variety of packaging options such as shrink wrap, modified atmosphere and printability allow plastic packaging to be tailored to the type of food to be contained. Thus, the food's visual appeal can be enhanced while maintaining preservation against ageing, degradation and chemical or microbial contamination.

Maintaining optical transparency in packaging is often desirable, especially from a customer's visual point of view and when observation of the package's contents is desirable for, for example, food or safety products. The optical properties of PHA can often be traded for the need to improve barrier properties over that of pure PHA homopolymer.[51] However, careful screening and characterisation of certain additives or fillers can produce improved barrier properties without compromising optical properties. For example, the addition of keratin derived from waste chicken feathers provided improved barrier properties to P(3HB) and P(HB-*co*-HV) without diminishing optical properties with a keratin content of up to 5%.[51]

Optical property optimisation has been known to be given an 'elaboration' or secondary priority status when packaging design has been considered. A far greater focus has been placed on ensuring good mechanical and barrier properties first with optical transparency, weld strength, hot tack and transport properties being optimised on a second pass.[52]

8.6.3 Thermal Properties

The thermal properties are of a vital consideration when selecting a polymer for packaging. Fortunately, PHAs provide (through diversity of structure and chemistry) a wide range to select for thermal properties suitable to packaging needs. Melting temperatures (T_m) from 60 to 177 °C, glass transition temperatures (T_g) from −50 to 4 °C and thermal degradation temperatures at highs of 256 to 277 °C are all within the range of PHAs currently being produced.[53] Despite this variety of thermal properties within the family of PHAs, it is important to note that P(3HB), the most commonly studied homopolymer, exhibits thermal instability during conventional melt processing at or near its melting point temperature (175 °C) and consequently provides only a narrow processing window for temperature driven processes.[54] The thermal degradation of PHB is understood to be due to a random chain scission process. The proximity of the melting point and thermal degradation temperature of P(3HB) have supported the inclusion of additives to separate these two temperatures and to better support thermal processing.

8.6.3.1 Additives to PHAs

Carboxyl-terminated butadiene acrylonitrile rubber (CTBA) or polyvinyl pyrrolidone (PVP) have been added to PHB in an effort to modify its thermal processing. Hong *et al.* report a significant modification of PHB crystallisation rate, crystallinity, melting temperature and thermal stability with the addition of only 1% (w/w) of these additives.[55] The improvement in thermal stability was proposed to be due to the steric hindrance that the PVP or CTBN exerted on the organisation of the PHB chains. Similarly, Wang and co-authors demonstrated how the addition of poly(*d,l*-lactide) (PDLLA) into P(3HB-*co*-3HV) substantially improved the thermal stability of this copolymer as measured with TGA and DSC. Furthermore, the addition of PEG to the blend of P(3HB-*co*-3HV) and PDLLA had the added benefit of accelerating degradation of the composite at room temperature, with a mass loss of 20% after 30 days.[56]

8.6.3.2 PHAs as Additives

In an effort to improve PHB's properties, Lim *et al.* synthesised a copolymer of P(3HB) with P(3HHx), P(3HB-*co*-3HHx), using fats and oils.[57] The properties of P(3HB-*co*-3HHx) extended the potential application range of pure PHAs and also provided another copolymer blend ingredient for other biopolymers such as PLA. Lim and co-authors demonstrated that increasing amounts of P(3HB-*co*-3HHx) melt blended with PLA increased suppression of PLA crystallization.[57] Despite a lack of chemical interaction, amorphous domains of P(3HB-*co*-3HHx) were found in the PLA and it was concluded to be the cause of a comparatively more ductile PLA with minimal aggregation of the P(3HB-*co*-3HHx).

8.6.4 Sustainability

Plastics derived from biological sources such as the PHA family of polymers are renowned for their environmentally friendly properties.[53] PHAs are biodegradable and sourced from the bacterial consumption of glucose or fatty acids; they not only contain zero percent petrochemical monomers but are a biological polymer created in an aqueous phase at temperatures of <40 °C.[53] Sustainability consists of environmentally favourable production, coupled with the environmental recycling of the product. Furthermore, the economic drive to reduce costs in PHA production has supported the use of carbon waste sources from agriculture and the sewage treatment system. Thus, waste carbon is given added value and effectively recycled, further supporting sustainability.

8.6.5 Barrier Properties

When compared to the other typical ff-polymers of polyethylene (PE) and polystyrene (PS) used in packaging, the oxygen and water barrier properties of commercial packaging PHA resins are considered to be naturally of a superior level (Figure 8.6).[58]

In order for the plastic packaging to be effective against aroma molecules, prevention of these molecules, often less than 400 MPa in size, from crossing the polymer barrier needs to be maintained.[58] The vapour pressure exerted by these small aroma bearing molecules provides an added challenge for the application of biopolymers in packaging. A packaging's vapour barrier properties relates to its ability to prevent water vapour from crossing the polymer packaging boundary. Several factors including mechanics, morphology and crystallinity can play a substantial role in determining a packaging's vapour barrier properties.

8.6.5.1 *Additives to PHAs for Barrier Enhancement*

One particular strategy of improving PHA packaging barrier properties would be to develop suitable nanocomposites. In particular, nanocomposites incorporating nanoclays of montmorillonite and kaolinite clays could also substantially improve the mechanical strength and thermal stability as well as the gas barrier properties. The addition of layered silicates or clay-like materials has also been demonstrated to improve the barrier properties of PHAs.[59] When an organically modified kaolinite and an organically modified montmorillonite (MMT) were incorporated into P(3HB) at 4 wt% loading, the oxygen permeability was found to decrease by 43% at 24 °C and a relatively humidity of 0%.[59]

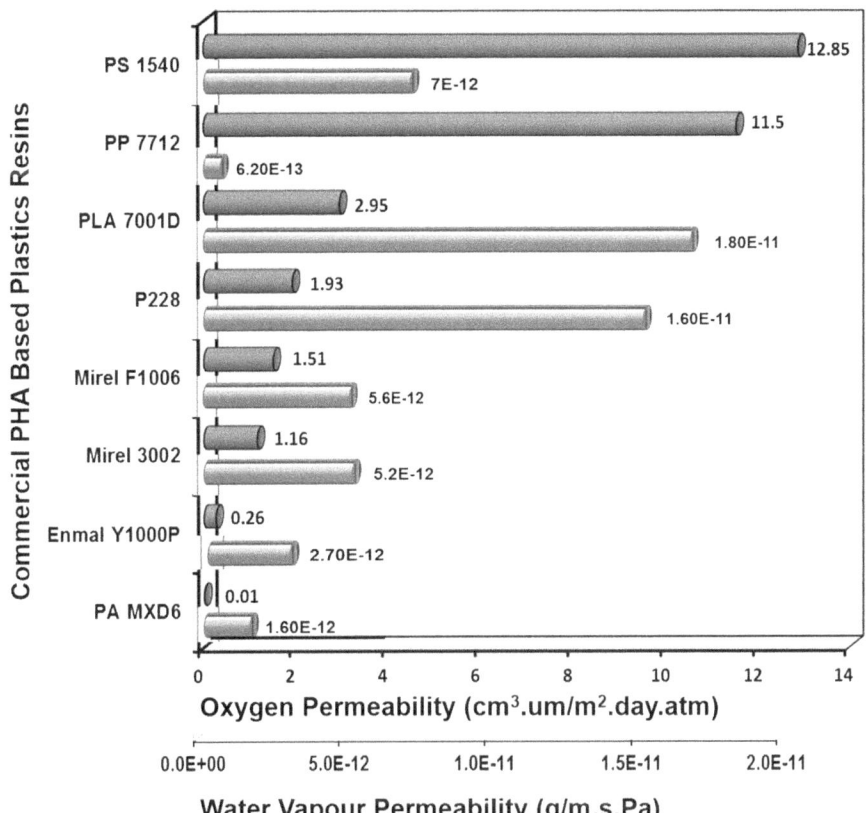

Figure 8.6 Barrier properties of commercial PHA resins compared to other bio- and oil-based polymers used for packaging, as measured by oxygen and water permeabilities, (PS1540: Polystyrene from Arkema; PP7712: Polypropylene from Total Chemical Company; PLA7001D: Poly(lactic acid) from Natureworks; P228: PHA from Biomer; MirelF1006 & 3002: PHA from Metabolix; Enmal Y1000P: PHA from Tianin Biologic; PA MXD6: PHA from Mitsubishi Chemical Company).
Adapted from Ref. 58.

Another commendable strategy of PHA barrier enhancement lies in the use of other biodegradable biopolymer additives. The seminal work involving adding keratin additives to PHAs through melt compounding using waste chicken feathers demonstrated how small quantities of additive (1 wt% keratin) added to P(3HB-*co*-3HV) could result in a significant reduction in water permeability of 59%, as well as a reduction in oxygen permeability of 67% and a reduction to limonene penetration (an organic aroma model compound) of 65%.[51]

8.6.5.2 PHAs as Additives for Barrier Enhancement

In a similar strategy to that used to improve the thermal properties of polymers, PHAs have been used as additives to improve the barrier properties of conventional, synthetic chemicals. For example, addition of P(3HB) to polyvinyl alcohol (PVOH) can lead to a significant improvement in its barrier properties. Thellen recently reported the use of a multilayer film strategy, where PVOH formed a core and PHA formed an outer skin layer. The barrier property of oxygen transmission rate (OTR) for this multilayered film was decreased by 55% at a relative humidity (RH) of 0% and showed a tremendous enhancement of 95% at a relative humidity of 60%.[60]

Using PHAs as an additive is only one strategy adopted to modify the water barrier properties of a material. A coating of P(3HB) on cellulose-based cardboard demonstrated significant improvements to its resistance to water uptake; a 90% reduction in moisture absorption was measured when the cardboard was coated with films containing 20% (w/w) P(3HB).[61]

Optimisation of the loading (wt%) of PHA additive is critical in ensuring maximum barrier conditions are obtained; non-optimisation can also lead to substantially decreased barrier properties, even lower than that exhibited by the unblended ff-polymer that the PHA additive is attempting to improve. The addition of P(3HB) to low density polyethylene (LDPE) was investigated by Pankova and co-authors for the purposes of drug delivery and biodegradation studies.[62] The authors determined that an optimal amount of 4.0 wt% addition of P(3HB) to LDPE produced a reciprocal water permeability of 7.3×10^{-9} (cm^2 h mm Hg)/(g cm). However, when the P(3HB) content was increased to 10 wt%, the reciprocal water permeability decreased significantly to 2.5×10^{-9} (cm^2 h mm Hg)/(g cm); this compares to 3.5×10^{-9} (cm^2 h mm Hg)/(g cm) for P(3HB) homopolymer.[62]

8.6.6 Degradation and Antimicrobial Properties

There are a wealth of studies on the degradation of PHAs, methodology of degradation and the microbial interaction with the polymer. The overall conclusion of these studies is that PHAs will degrade in practically any natural environment when exposed to a microbial community, although the rates of this degradation are dependent upon the environmental conditions. In this section, we demonstrate the potential importance of these properties in the development of plastic packaging using a few articles that deal directly with the topic of preventing microorganisms from crossing or establishing colonies on PHA-based packaging surfaces. Similarly, the antimicrobial properties directly correlate with the degradation rates and behaviour of PHA-based packaging systems.[63]

Unlike petroleum-based polymers, PHAs are broken down in the environment and the rate of degradation is influenced by factors such as pH, moisture, temperature, the molecular weight of both the individual's monomer units and the chain lengths as well as crystallinity and the absence or presence of antimicrobials within the PHA. The degradation of PHA packaging is considered to be the natural consumption of the polymer by biological microorganisms to produce CO_2 and H_2O.

Figure 8.7 Photographs illustrating material loss of a Mirel™ PHA plastic bottle cap through biodegradation when exposed to environmental conditions (images supplied by Mirel™).

In a study by Bucci *et al.,* P(3HB) packaging in the form of a jar and cap were fabricated through injection moulding, and both the packaging and degradation properties of the P(3HB) were investigated. Degradation of the jar and cap was examined in several media including sewage residues rich in organic material. Complete degradation of the P(3HB) packaging in these media was of the order of 90 days with no degradation observed in the control of potable water only; illustrating a useful result for short term water storage options.[64] Figure 8.7 shows the ready environmental degradation of a bottle cap produced using Mirel® PHA resin, which contrasts with the persistence displayed by conventional petroleum-based thermoplastics (Figure 8.7).

A shortcoming highlighted in the studies by Bucci and coworkers was the difficulties associated with successful injection moulding of P(3HB). The defect rate of the jar and cap packages was approximately 72%; test samples were deemed defective through cracking or incomplete injection. Furthermore, P(3HB) has been observed to yield an odour intensity significantly greater than PP. In addition, its tainting of water, as noted by taste testers, suggests that P(3HB) is unlikely to be an alternative plastic for water containment.[65]

8.7 PHAs for Food and Industrial Packaging

While P(3HB) is unsuitable for injection moulding, its copolymers have been successfully processed into commercial packaging items. For example, commercial grade PHAs such as Mirel's P(3HB-*co*-4HB), Tianjin's P(3HB-*co*-4HB) and Tianan's P(3HB-*co*-3HV) have been used to blow mould bottles for food packaging.[66] Furthermore, the potential application of comparatively more flexible medium chain length PHAs in food packaging has also been explored. For example, the potential application of mclPHAs as a film for cheese coating was reported by van der Walle *et al.* where the mclPHA film exhibited superior properties of low water vapour permeability compared to other ff-polymers, and consequently moisture loss was minimised.[67]

While PHA-based resins have been explored for the containment of industrial products and prevention of damage during transport, research has also focused on utilising their biodegradability to deliver chemical agents. Prudnikova *et al.* have investigated a blend of P(3HB-*co*-3HV) and poly(ε-caprolactone), (PCL), as an outer package casing for the delivery of the herbicide Zellek Super™. The herbicide product was successful delivered over a period of time with gradual degradation of the PHA packaging and was successful in controlling the growth of creeping bentgrass, *Agrostis stolonifera*.[68] Notably, adjustment of the content of the PHA casing enabled a level of control over degradation rate and product release.

8.8 PHAs for Medical Packaging

Similar to the industrial packaging for the delivery of agricultural chemical agents, medical packaging uses of PHAs remains primarily at the research stage and also focuses on drug delivery systems. As PHA biomaterials can be used to package the 'product' (drug) and deliver it to the 'user' (tissue), pharmaceutical drug delivery can be considered as a type of packaging according to its definition.

8.8.1 Biocompatibility of PHAs

Both P(3HB) and its monomeric component, 3-hydroxybutyric acid (3-HBA), are produced in mammalian systems.[69] In the body, low molecular weight P(3HB) forms complexes with other biomacromolecules, which significantly modifies its physico–chemical properties, permitting it to pervade both hydrophobic and aqueous regions of the cell. Thus, complexed P(3HB) is found in the cytoplasm and serves a transport function in cell membranes. In contrast, 3-HBA is an essential, stable analyte in human sera and one of the three ketone bodies produced during prolonged starvation and diabetes.[69] Similarly, 4-hydroxybutanoic acid, the monomeric component of P(4HB), is also widely distributed in the mammalian body.[70] The biocompatibility of P(3HB) and its breakdown products led, in part, to its initial commercialisation by W. R. Grace and Co. for medical devices in the late 1950s, while 4-HBA has long been utilised as an intravenous agent for long term sedation and for the induction of anesthesia.[71,72]

The biocompatibility and degradability of P(3HB), P(3HB-*co*-3HV), P(3HB-*co*-3HHx) and P(4HB) have been well documented.[73,74] Doyle *et al.* implanted injection moulded P(3HB) plaques and hydroxyapatite/PHB composites into femoral defects in New Zealand White rabbits; 12 months later no implant resorption had occurred while new bone was formed adjacent to the implant, with up to 80% of the implant's surface lying in direct apposition to new bone.[75] Löbler *et al.* tested PHB patches for the gastrointestinal tract in a rat model.[76] Asymmetric PHB patches were sutured onto the stomach wall and failed to induce fibrosis or a strong inflammatory response.[76]

P(3HB-*co*-3HV) has been investigated for potential use in a diverse range of medical applications including bone repair implants, medical sutures, cardiovascular stents and as drug carriers.[77,78] Wu *et al.* have reported that solvent cast P(3HB-*co*-3HV) films fabricated from commercial supply were contaminated with lipopolysaccharide (LPS).[79] The authors report that these 'unpurified' films induced macrophage activation resulting in expression of pro-inflammatory tumour necrosis factor (TNF) and interleukin-6 (IL-6) during *in vitro* studies.[79] However, Foster and coworkers applied a cell cycle approach to investigate the influence of the biomaterial on olfactory ensheathing cells (OECs) and mesenchymal stem cells (MSCs) and found that significantly higher percentages of cells were cycled at the synthesis (S) phase of the cell cycle when cultivated on P(3HB-*co*-3HV) films compared to P(3HB), with MSCs more susceptible than OECs.[80,81] Furthermore, Chaput *et al.* studied P(3HB-*co*-3HV) patches, with various loadings of hydroxyvalerate, implanted intramuscularly in sheep for a lengthy 90 weeks.[82] After an initial inflammatory response observed at one week, the reaction decreased. After 11 weeks, the capsule at the interface between the polymers and muscular tissue consisted primarily of connective tissue cells that were well vascularised with highly organised, orientated fibres. In addition, fibroblastic cells were observed, aligned in parallel with the biopolymer surface.[82]

8.8.2 PHAs for Drug Delivery

The properties of PHA biocompatibility and degradability in mammalian systems have been utilised for packaging of therapeutic agents that can subsequently serve as drug delivery agents and carriers in blood.[83] The challenges facing the effective application of PHAs for drug delivery include (a) targeted delivery and (b) controlled or sustained release of the active drug to the target region. A drug 'packaged' using a PHA-based carrier would take advantage of the degradation rate of the PHA in order to effect sustained delivery of the drug payload. The breakdown of PHAs in mammalian systems is predominantly through chemical hydrolysis where crystallinity and molecular weight play important roles. Therefore, whilst carriers produced from sclPHAs, such as PHB and P(3HB-*co*-3HV), are considered porous and highly crystalline (properties that would lend a carrier to deliver its payload too quickly), mclPHAs, such as polyhydroxyoctanoate (PHO), being comparatively more amorphous with a lower crystallinity and melting point, appear more suited to drug delivery.[83]

Various forms of PHAs have been used to package drugs and control their delivery, including gels, films, microcapsules, microspheres, nanoparticles, porous matrices, polymeric micelles and polymer linked drugs.[84] For example, Kassab *et al.* used a simple solvent evaporation technique to produce P(3HB) microspheres packaging Rifampicin as a chemoembolising agent, with 90% of the drug released within 24 hours.[85] Similarly, Turesin *et al.* compared P(3HB), P(3HB-*co*-3HV) and P(3HB-*co*-4HB) as rods for the release of the antibiotics Sulperazone® and Duocid® and reported that their release was controlled by the PHA to drug ratio. Sulperazone® dissolution was significantly greater than the degradation of the PHA carrier.[86]

8.8.3   PHAs for Cell Delivery

PHAs have attracted considerable attention as scaffolds to deliver cells to the body, in order to support tissue regeneration and engineering.[74,87] For example, P(3HB) fibres have been used as scaffolds to deliver matrix components and cell lines supporting neuronal survival and regeneration after spinal cord injury. Novikov *et al.* constructed a scaffold consisting of P(3HB) fibres coated with fibronectin and alginate hydrogel.[88] Fibres were implanted in the lesion cavity after cervical spinal cord injury in adult rats with neurons of the rubrospinal tract used as an experimental model. In untreated animals, 45% of the injured neurons were lost 8 weeks postoperatively. In contrast, the P(3HB) fibre scaffolds reduced this loss to approximately 23%, a similar effect to treating the wound with neurotrophic factors BDNF or NT-3.[88] The authors also used these P(3HB) scaffolds to deliver neonatal Schwann cells prior to implantation and this resulted in regenerating axons entering the fibre from both ends and extending along its entire length. Thus, PHB-based fibre scaffolds were successfully used to deliver Schwann cells to support neuronal survival and regeneration after spinal cord injury.[88]

Opitz *et al.* implanted P(4HB) scaffolds seeded with autologous vascular smooth muscle cells (vSMC) and ovine endothelial cells from carotid arteries, for 14 days in the descending aorta of juvenile sheep.[89] These constructs were implanted after *in vitro* endothelialisation and periodically examined up to 24 weeks for evaluation. The authors report that up to three months after implantation, the grafts were fully patent, with no signs of dilatation, occlusion or intimal thickening. In addition, a confluent luminal endothelial cell layer was observed. In contrast, the six-month grafts displayed significant dilatation and partial thrombus formation while histology displayed layered tissue formation resembling native aorta.[89] Similarly, Hoerstrup *et al.* have used PHO to fabricate trileaflet heart valves and seeded these with vascular cells harvested from ovine carotid arteries.[90] After long term (152 weeks) implantation in lambs, these tissue-engineered constructs were eventually covered with tissue, with no thrombus formation on any of the specimens.[91]

8.9 Conclusions and Future Work

Packaging serves a number of essential functions, preventing physical and biological damage to goods. While the material versatility of plastics has led to their dominance in the packaging industry, their environmental persistence coupled with environmental and public health concerns regarding their various additives, is recognised as a serious issue. Consequently, there is considerable interest in natural, environmentally-friendly alternatives to conventional plastics synthetically produced from depleting fossil fuels. In addition, natural, biodegradable, biological plastics, or 'bioplastics', would support sustainable development by promoting the natural recycling of these materials. Through its diversity, the PHA family offers versatility in the properties required for plastic packaging. However, economically viable production has limited their commercial progress to those consisting of short chain lengths.

Compared to current plastics used for packaging, PHA-based bioplastics also exhibit biodegradability. While their biodegradation ensures the continued development of PHAs for sustainable packaging, researchers are also using this property for the controlled release of agricultural agents such as herbicides and pesticides. Similarly, the biocompatibility of many PHAs, together with FDA approval for PHB, have supported their investigation in the packaging and delivery of pharmaceutical agents and cells in a variety of medical applications. Thus, PHAs have been used to produce value-added packaging with properties beyond current plastics while promoting an environmentally-friendly alternative for a sustainable future.

Online shopping and globalisation of markets is generating a growing increase in the use of plastic packaging as well as an awareness of their shortcomings. Thus the drive for sustainable packaging will see an increase in the use of PHAs both as market share and application diversity. Furthermore, the unique properties of biodegradability and biocompatibility will see the continued development of PHA-based, bioplastic packaging of biologically active agents for delivery systems.

References

1. Northern Ireland Environment Agency. http://www.doeni.gov.uk/niea/waste-home/regulation/regulations_packaging/definition_of_packaging.htm (accessed 21/01/2014).
2. US Environmental Protection Agency http://www.epa.gov/osw/conserve/materials/plastics.htm (accessed 21/01/2014).
3. L. M. Rios, C. Moore and P. R. Jones, *Mar. Pollut. Bull.*, 2007, **54**, 1230.
4. R. C. Thompson, S. H. Swan, C. J. Moore and F. S. vom Saal, *Philos. Trans. R. Soc., B*, 2009, **364**, 1973.
5. A. L. Andrady, *Mar. Pollut. Bull.*, 2011, **62**, 1596.
6. Plastics Europe (2010), Plastics – The Facts 2010.

7. World Packaging Organisation Market Statistics and Future Trends in Global Packaging PIRA International Ltd (2008).
8. European Commission, Eurostat, packaging waste statistics http://epp.eurostat.ec.europa.eu/statistics_explained/index.php/Packaging_waste_statistics (accessed 21/01/2014).
9. D. K. A. Barnes, F. Galgani, R. C. Thompson and M. Barlaz, *Philos. Trans. R. Soc. B*, 2009, **364**, 1985.
10. C. J. Moore, *Environ. Res.*, 2008, **108**, 131.
11. P. G. Ryan, C. J. Moore, J. A. Van Franeker and C. L. Moloney, *Philos. Trans. R. Soc. B*, 2009, **364**, 1999.
12. R. C. Thompson, *Marine Nature Conservation in Europe*, ed. J. C. Krause, H. Nordheim and S. Bräger, Federal Agency for Nature Conservation, Stralsund, Germany, 2006, pp. 107–115.
13. Australian Bureau of Statistics, *Document 4613.0 Australia' Environment: Issues and Trends*, Jan 2010.
14. The Packaging Council of Australia, *Packaging: The Statistics*, www.pca.org.au/uploads/00207.pdf (accessed 02/04/2014).
15. S. Modi, K. Koelling and Y. Vodovotz, *Eur. Polym. J.*, 2011, **47**, 179.
16. J. A. Colwill, E. I. Wright and S. Rahimifard, *J. Polym. Environ.*, 2012, **20**, 1112.
17. P. K. Roy, M. Hakkarainen, I. K. Varma and A.-C. Albertsson, *Environ. Sci. Technol.*, 2011, **45**, 4217.
18. M. A. Browne, T. Galloway and R. Thompson, *Integr. Environ. Assess. Manage.*, 2007, **3**, 559.
19. Å. Lithner, G. Larsson and G. Dave, *Sci. Total Environ.*, 2011, **409**, 3309.
20. E. L. Teuten, J. M. Saquing, D. R. U. Knappe, M. A. Barlaz, S. Jonsson, A. BjÃrn, S. J. Rowland, R. C. Thompson, T. S. Galloway, R. Yamashita, D. Ochi, Y. Watanuki, C. Moore, P. H. Viet, T. S. Tana, M. Prudente, R. Boonyatumanond, M. P. Zakaria, K. Akkhavong, Y. Ogata, H. Hirai, S. Iwasa, K. Mizukawa, Y. Hagino, A. Imamura, M. Saha and H. Takada, *Philos. Trans. R. Soc. B*, 2009, **364**, 2027.
21. J. r. Oehlmann, U. Schulte-Oehlmann, W. Kloas, O. Jagnytsch, I. Lutz, K. O. Kusk, L. Wollenberger, E. M. Santos, G. C. Paull, K. J. W. Van Look and C. R. Tyler, *Philos. Trans. R. Soc. B*, 2009, **364**, 2047.
22. F. S. vom Saal and J. P. Myers, *J. Am. Med. Assoc.*, 2008, **300**, 1353.
23. T. Colborn, D. Dumanoski and J. P. Myers, *Our stolen future: are we threatening our fertility, intelligence and survival? A scientific detective story*, Dutton, New York, 2006.
24. A. K. Hotchkiss, C. V. Rider, C. R. Blystone, V. S. Wilson, P. C. Hartig, G. T. Ankley, P. M. Foster, C. L. Gray and L. E. Gray, *Toxicol. Sci.*, 2008, **105**, 235.
25. S. H. Safe, *Trends Endocrinol. Metab.*, 2005, **16**, 139.
26. R. H. Waring and R. M. Harris, *Mol. Cell. Endocrinol.*, 2005, **244**, 2.
27. S. H. Swan, K. M. Main, F. Liu, S. L. Stewart, R. L. Kruse, A. M. Calafat, C. S. Mao, J. B. Redmon, C. L. Temand, S. Sullivan and J. L. Teague, *Environ. Health Perspect.*, 2005, **113**, 1056.

28. K. L. Howdeshell, A. K. Hotchkiss, K. A. Thayer, J. G. Vandenbergh and F. S. vom Saal, *Nature*, 1999, **401**, 763.
29. I. Colón, D. Caro, C. J. Bourdony and O. Rosario, *Environ. Health Perspect.*, 2000, **108**, 895.
30. M. D. Anyway, A. S. Cupp, M. Uzumcu and M. K. Skinner, *Science*, 2005, **308**, 1466.
31. International Standards Organisation, http://www.iso.org/iso/home.html (accessed 02/05/2014).
32. L. J. R. Foster, R. W. Lenz and R. C. Fuller, *Hydrogels and Biodegradable Polymers For Bioapplications*, ed. R. Ottenbrite, K. Park and S. Huang, ACS Books, Washington D.C., USA, 1996, pp. 68–92.
33. E. A. Dawes and P. J. Senior, *Adv. Microb. Physiol.*, 1973, **10**, 135.
34. L. J. R. Foster, *Appl. Microbiol. Biotechnol.*, 2007, **75**, 1241.
35. B. Hazer and A. Steinbüchel, *Appl. Microbiol. Biotechnol.*, 2007, **74**, 1.
36. G. Braunegg, G. Lefebvre and K. F. Genser, *J. Biotechnol.*, 1998, **65**, 127.
37. M. H. Deinema, *Appl. Microbiol.*, 1972, **24**, 857.
38. L. L. Wallen and W. K. Rohwedder, *Environ. Sci. Technol.*, 1974, **8**, 576.
39. G. Odham, A. Tunlid, G. Westerdahl and P. Marden, *Appl. Environ. Microbiol.*, 1986, **52**, 905.
40. J. Mas-Castella and R. Guerrero, *Can. J. Microbiol.*, 1995, **41**, 80.
41. J. S. Nickels, J. D. King and D. C. White, *Appl. Environ. Microbiol.*, 1979, **37**, 459.
42. R. H. Findlay and D. C. White, *Appl. Environ. Microbiol.*, 1983, **45**, 71.
43. L. J. R. Foster, A. Saufi and P. J. Holden, *Biotechnol. Lett.*, 2001, **23**, 1519.
44. A. Khan, T. Huq, R. A. Khan, B. Riedl and M. Lacroix, *Crit. Rev. Food Sci. Nutr.*, 2014, **54**, 163.
45. M. S. Sreekanth, S. V. N. Vijayendra, G. J. Joshi and T. R. Shamala, *J. Food Sci. Technol.*, 2013, **50**, 404.
46. Q. Chen and L. H. Zhang, *Appl. Mech. Mater.*, 2014, **448**, 160.
47. A. M. Gumel, M. S. M. Annuar and Y. Chisti, *J. Polym. Environ.*, 2013, **21**, 580.
48. S. Gibon, V. Arondel, K. Iba and C. Somerville, *Am. Soc. Plant Biol.*, 1994, **106**, 1615.
49. H. M. C. Azeredo, *Food Res. Int.*, 2009, **42**, 1240.
50. E. Ohashi, W. S. Drumond, N. P. Zane, P. W. D. F. Barros, M. G. Lachtermacher, H. Wiebeck and S. H. Wang, *Macromol. Symp.*, 2009, **279**, 138.
51. P. Pardo-Ibáñez, A. Lopez-Rubio, M. Martínez-Sanz, L. Cabedo and J. M. Lagaron, *J. Appl. Polym. Sci.*, 2014, **131**, 39947.
52. P. Bordes, E. Pollet and L. Averous, *Prog. Polym. Sci.*, 2009, **34**, 125.
53. G.-Q. Chen and M. K. Patel, *Chem. Rev.*, 2012, **112**, 2082.
54. C. Johansson, J. Bras, I. Mondragon, P. Nechita, D. Plackett, P. Šimon, D. G. Svetec, S. Virtanen, M. G. Baschetti, C. Breen, F. Clegg and S. Aucejo, *BioResources*, 2012, **7**, 2506.
55. S.-H. Hong, T.-S. Gau and S.-C. Huang, *J. Thermal Anal. Calorim.*, 2011, **103**, 967.

56. S. Wang, P. Ma, R. Wang and Y. Zhang, *Polym. Degrad. Stab.*, 2008, **93**, 1364.

57. J. S. Lim, K. I. Park, G. S. Chung and J. H. Kim, *Mater. Sci. Eng. C.*, 2013, **33**, 2131.

58. M. Vähä-Nissi, T. Hjelt, M. Jokio, R. Kokkonen, J. Kukkonen and A. Mikkelson, *Packag. Technol. Sci.*, 2008, **21**, 425.

59. M. D. Sanchez-Garcia, E. Gimenez and J. M. Lagaron, *J. Appl. Polym. Sci.*, 2008, **108**, 2787.

60. C. Thellen, S. Cheney and J. A. Ratto, *J. Appl. Polym. Sci.*, 2013, **127**, 2314.

61. V. P. Cyras, C. M. Soledad and V. Analía, *Polymer*, 2009, **50**, 6274.

62. Y. N. Pankova, A. N. Shchegolikhin, A. L. Iordanskii, A. L. Zhulkina, A. A. Olkhov and G. E. Zaikov, *J. Mol. Liq.*, 2010, **156**, 65.

63. C. Woolnough, L. Yee, T. Charlton and L. J. R. Foster, *Polym. Int.*, 2010, **59**, 658.

64. D. Z. Bucci, L. B. B. Tavares and I. Sell, *Polym. Test.*, 2007, **26**, 908.

65. D. Z. Bucci, L. B. B. Tavares and I. Sell, *Polym. Test.*, 2005, **24**, 564.

66. J. Greene, *Plast. Eng.*, 2013, **69**, 16.

67. G. A. van der Walle, G. J. de Koning, R. A. Weusthuis and G. Eggink, *Adv. Biochem. Eng. Biotechnol.*, 2001, **71**, 263.

68. V. Prudnikova, A. N. Boyandin, G. S. Kalacheva and A. J. Sinskey, *J. Polym. Environ.*, 2013, **21**, 675.

69. L. J. R. Foster and B. J. Tighe, *J. Environ. Polym. Degrad.*, 1994, **2**, 185.

70. D. P. Martin and S. F. Wlliams, *Biopolymers*, 2003, **16**, 97.

71. L. J. R. Foster, *Polymers from Renewable Resources: Biopolyesters and Biocatalysis*, ed. C. Scholz and R. A. Gross, ACS Books, Washington D.C., USA, 1996, pp. 42–66.

72. E. Entholzner, L. Mielke, R. Pichlmeier, F. Weber and H. Schneck, *Der Anaesthesist*, 1995, **44**, 345.

73. M. Unverdorben, A. Spielberger, M. Schywalsky, D. Labahn, S. Hartwig, M. Schneider, D. Lootz, D. Behrend, K. Schmitz, R. Degenhardt, M. Schaldach and C. Vallbracht, *Cardiovasc. Intervent. Radiol.*, 2002, **25**, 127.

74. G.-Q. Chen and Q. Wu, *Biomaterials*, 2005, **26**, 6565.

75. C. Doyle, E. T. Tanner and W. Bonfield, *Biomaterials*, 1991, **12**, 841.

76. M. Löbler, M. Sax, K. P. Schmitz and U. T. Hopt, *J. Biomed. Mater. Res.*, 2003, **61**, 165.

77. E. I. Shishatskaya, T. G. Volova, A. P. Puzyr, O. A. Mogilnaya and S. N. Efremov, *J. Mater. Sci. Mater. Med.*, 2004, **15**, 719.

78. E. Shishatskaya, I. A. Khlusov and T. G. Volova, *J. Biomater. Sci., Polym. Ed.*, 2006, **17**, 481.

79. A. C. Wu, L. Grondahl, K. S. Jack, M. X. Foo, M. Trau, D. A. Hume and A. I. Cassady, *Biomaterials*, 2006, **27**, 4715.

80. R. T. H. Chan, R. A. Russell, H. Marcal, T. H. Lee, P. J. Holden and L. J. R. Foster, *BioMed. Res. Int.*, 2014, **2014**, 676493.

81. T. Ahmed, H. Marcal, M. Lawless, N. Wanandy, A. Chiu and L. J. R. Foster, *Biomacromolecules*, 2010, **11**, 2707.

82. C. Chaput, L. H. Yahia, A. Selmani and C.-H. Rivard, *Mater. Res. Soc. Symp. Proc.*, 2005, **385**, 49.
83. D. B. Hazer, E. Kılıçay and B. Hazer, *Mater. Sci. Eng. C.*, 2012, **32**, 637.
84. H. Ueda and Y. Tabata, *Adv. Drug Delivery Rev.*, 2003, **55**, 501.
85. A. C. Kassab, K. Xu, E. B. Denkbas, Y. Dou, S. Zhao and E. Piskin, *J. Biomater. Sci., Polym. Ed.*, 1997, **8**, 947.
86. F. Turesin, I. Gursel and V. Hasirci, *J. Biomater. Sci., Polym. Ed.*, 2001, **12**, 195.
87. G. T. Köse, F. Korkusuz, P. Korkusuz and V. Hasirci, *Tissue Eng.*, 2004, **10**, 1234.
88. L. N. Novikov, A. Mosahebi, M. Wiberg, G. Terenghi and J. O. Kellerth, *Biomaterials*, 2002, **23**, 3369.
89. F. Opitz, K. Schenke-Layland, W. Richter, D. P. Martin, I. Degenkolbe, T. Wahlers and U. A. Stock, *Ann. Biomed. Eng.*, 2004, **32**, 212.
90. S. P. Hoerstrup, R. Sodian, S. Daebritz, J. Wang, E. A. Bacha, D. P. Martin, A. M. Moran, K. J. Guleserian, J. S. Sperling, S. Kaushal, J. P. Vacanti, F. J. Schoen and J. E. Mayer, *Circulation*, 2000, **102**, 44.
91. R. Sodian, S. P. Hoerstrup, J. S. Sperling, S. Daebritz, D. P. Martin, A. M. Moran, B. Y. Kim, F. J. Schoen, J. P. Vacanti and J. E. Mayer, *Circulation*, 2000, **102**, 22.

CHAPTER 9

Packaging Applications of Polyhydroxyalkanoates (PHAs)

FARAYDE MATTA FAKHOURI,[*][a,b] MARCELO CARVALHO,[b] PEDRO LUIS MANIQUE BARRETO,[c] RODOLFO CARDOSO DE JESUS[d] AND SILVIA MARIA MARTELLI[*][a]

[a] Faculty of Engineering, Federal University of Grande Dourados (UFGD), Dourados, Brazil; [b] Department of Chemical Engineering, Campinas State University (UNICAMP), Campinas, Brazil; [c] Department of Food Science and Technology, Centre for Agricultural Sciences, Federal University of Santa Catarina (UFSC), Florianópolis, Brazil; [d] Faculty of Technology of São Paulo State (FATEC), Mogi-Mirim, Brazil
*Email: farayde@gmail.com; smmartelli@gmail.com

9.1 Introduction

Over the past few decades, a new food packaging concept has been studied, combining the areas of food technology, biotechnology, materials science and packaging.[1] In this context, some new concepts have been widely publicized in research areas such as the term biopolymer, a special type of polymer that involves living organisms in their synthesis[2] with partial or full biochemical origin, which may be partially or completely made from natural materials, are renewable (biomass) and may be biodegradable, meeting ASTM standards,[3] or are non-biodegradable.[4]

Bioplastics are defined as materials containing biopolymers in variable amounts. They can be shaped by heating and pressure. They are a potential

RSC Green Chemistry No. 30
Polyhydroxyalkanoate (PHA) Based Blends, Composites and Nanocomposites
Edited by Ipsita Roy and Visakh P M

alternative to conventional thermoplastic polymers (petrochemical origin), such as polyolefins and polyesters.[4]

Biopolymers have been widely used in the packaging industry as substitutes for conventional plastic packages. According to these authors, biopolymers can be classified into:

(i) Polymers derived from biomass, with or without modification (starch).
(ii) Polymers produced by microrganisms in their natural state or genetically modified (PHA, polyhydroxyalkanoates).
(iii) Polymers produced with bio-intermediate participation. These bio-intermediates are produced from renewable sources, such as PLA (poly lactic acid).

There is not an international consensus about the content of bio-based polymer required to qualify the product as a bioplastic yet. The percentage of renewable carbon source that has been debated is between 25 and 40%.[4]

One of the most versatile bioplastics is PLA, a type of polyester obtained by lactic acid polymerization, resulting from sugar fermentation. Besides its use in disposable packaging (Ingeo) that is already in use by several companies (Coca-Cola, McDonald's), it is also used as pillow filling and in comforters (NatureWorks), in film and paper coating (BASF) and it is also in use in the automotive (Hyundai) and electronic industries (Samsung).

Polymers synthesized by microorganisms directly, such as polyhydroxyalkanoates (PHAs) and poly-hydroxybutyrate (PHB), are entering the food packaging market slowly. Due to their biocompatibility, they have also found important applications in the field of medicine (Biopol produced by Monsanto is made of a copolymer composed of PHB).

It is worth noting that biopolymers made from sugar represent a promising business; once they are produced using clean technology, the process will aid in preserving the environment, and it is able to be integrated into the existing production process in plants. Due to the sugarcane production cycle, plant devices are traditionally idle for at least half a year.[5]

9.2 Literature Review

The international oil price rise, geopolitical instability in important oil production regions and the global need to develop technologies with low greenhouse gas emissions have all pointed to the real possibility of exploiting raw materials such as renewable feedstocks for intermediate biosynthesis and final chemical products. According to Silva *et al.*,[6] the experience of using ethanol as fuel and the incentive for biodiesel production are factual examples of using renewable raw materials to replace those derived from petroleum. Besides the energy supply needs, the second largest petroleum demand is for its use as raw material for polymer production.

Petroleum-based polymers are used for different purposes and there is a huge scale to produce these polymers. However, the increase in the price of oil barrels and the increase of the global population's awareness of environmental issues (larger volumes of all kind of plastics in landfills, global temperatures rising and glaciers melting) have created a propitious scenario to produce biopolymers on a larger scale.

Polymer use from renewable sources has been widely studied in the biomedical field and in the packaging industries. Natural polymers are generally biodegradable, biocompatible and can be obtained at relatively low cost.[7]

A wide range of biodegradable alternatives have been proposed to mitigate the problem.[6] The raw materials used can be fully renewable, such as bacterial polyhydroxyalkanoates (PHAs) and polylactide (PLA) proposed by Dow-Cargill consortium, or non-renewable materials, such as alternatives from Showa Highpolymer (Japan) and BASF. The main precursors for the synthesis of the latter polymers may be potentially produced through microbes in the near future, using renewable raw materials. Bacterial polyhydroxyalkanoates are a polymer class studied extensively and they have biodegradability characteristics besides being renewable. Silva *et al.* (2007) also state that two components are crucial for biopolymers to become competitive in the packaging market: the cost of raw materials and energy and the capital investment needed to build an industrial production plant.

Once biopolymers are biodegradable, they are well regarded by environmentalists. When compared to traditional polymers, biopolymers are derived from a renewable carbon source (sugarcane, corn, potato, wheat and sugar beet, or a vegetable oil extracted from soybeans, sunflower, palm or other oleaginous plant). Despite this advantage of biopolymers compared to traditional polymers, anaerobic biodegradation of many natural biopolymers can lead to methane and hydrogen sulfide (H_2S) formation. Methane is a stronger greenhouse gas than carbon dioxide (CO_2) and it is slower to be reabsorbed by natural processes than CO_2.

Biodegradable materials do not present a risk of environmental impact if these materials can biodegrade completely within six months. The rule was established, but this does not usually occur with many biopolymers. The CO_2 resulting from burning or biodegradation of these materials is renewable and enters the carbon mass balance in the environment, however, methane and other gases can accumulate easily. This shows us that using biodegradable materials does not stop the need for rational and conscious usage, focused on recycling, reuse and rational disposal.

Biopolymers, however, present numerous advantages: (i) they are biodegradable, (ii) they are biocompatible (they can be used on the human body as prostheses, implants *etc.*), (iii) they can be produced by some effluents and industrial by-products, particularly from the food industry, (iv) biopolymer production takes into account environmental issues, (v) they have a wide range of applications and properties, (vi) production costs have been decreasing as a result of the current interest from the environmental area and new technology is available, (vii) specific applications in biomaterials and nanotechnology areas have been causing a rise in the commercial value of biopolymers.

With the biopolymers that have been studied since 1991, there has been an incentive for studying the production of biodegradable plastics with diverse approaches. Great efforts have resulted in the development of the production processes of poly(hydroxybutyrate) (PHB) and poly(3-hydroxybutyrate-*co*-3-hydroxyvalerate) (PHB-*co*-HV), two polymers from the polyhydroxyalkanoates family, from sucrose as a primary carbon source. The technology was transferred to PHB Industrial S/A company, following different studies for the characterization and application of PHAs.

PHA polymers are synthesized by a wide variety of bacterial strains with intracellular carbon and energy accumulation under adverse growth conditions and in the presence of abundant carbon sources. *Ralstonia eutropha* (or *Alcaligenes eutrophus*) is one of the most studied microorganisms for PHA production due to its ease of growth using renewable carbon sources and the fact that this bacteria may reach up to 80% of its dry weight as a polymer.[8]

9.2.1 Polyhydroxyalkanoates (PHAs)

The term PHA (Figure 9.1) is applied to a family of polyesters accumulated by various bacteria, deposited in the cells in the form of highly refractive granules of carbon reserve, energy and reducing equivalents.[7-9] In general, PHA synthesis by bacteria in a nutrient medium occurs when there is an excess of carbon source and a lack of one necessary nutrient at least (N, P, Mg, Fe *etc.*) for cell multiplication.[6]

The polyhydroxyalkanoate is a fully biodegradable polyester in microbiologically active environments. It can be synthesized by modified plants or bacteria from intracellular reserve materials *via* direct biosynthesis of carbohydrates from sugarcane, corn, or vegetable oils mainly extracted from soybeans and palm. PHA production has several advantages: biodegradability, a decrease of environmental impact, replacement of conventional plastics and the use of new materials in fields such as medicine. On the other hand, there is a disadvantage in that PHA production still involves high costs.

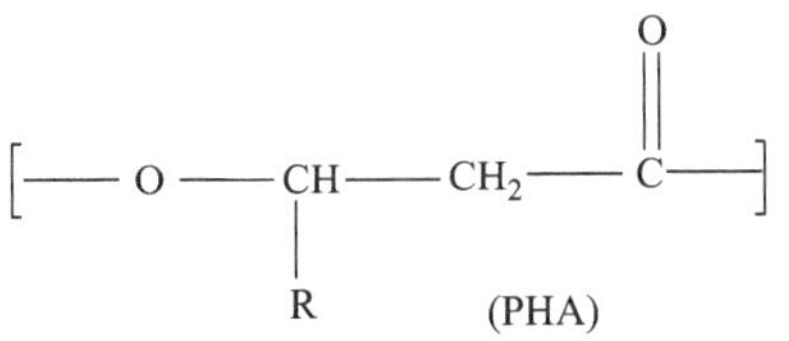

Where R = methyl or ethyl (short chain length); R = propyl, butyl, pentyl, hexyl or heptyl (medium chain length); R = more than 14 carbons per repeating unit (long chain length).

Figure 9.1 PHA generic monomer.

PHAs are made from renewable sources such as sugarcane, one of the most significant and traditional Brazilian feedstocks, as well as from other renewable sources such as starch, vegetable oils *etc.* Their applications range from packaging to disposables and applications in the medical field (matrices for slow drug release, implants, artificial tissues, patterns and suture wire) due to their biocompatibility. These biopolymers are similar to plastics produced from petroleum and can be laminated, molded and injected easily. Also, they are fully biodegradable in a period of six months to a year.[9]

Polyhydroxyalkanoates are polyesters accumulated as energy storage materials by various bacteria in the form of intracellular granules, reaching up to 80% of bacterial biomass weight. They can be synthesized by different sources of renewable and non-renewable carbon, such as molasses (from sugar cane) and even some effluents from the food industry. Nowadays, PHAs have been obtained from bacteria and some genetically modified plants.

Bacterial PHA biosynthesis takes place in bioreactors where there is an excess of carbon source and a lack of one nutrient, at least, necessary for cell multiplication (N, P, Mg, S, Fe, O_2, K, Cu, Co and Zn). The polymer is accumulated within bacterial cells as granules, reaching 80–90% of the cells' dry weight. PHA's composition and its molecular weight depend on the chemical nature of the raw material offered as a carbon source, the bioreactor's environmental operating conditions and the type of bacteria inoculated. Thus, polymer characteristics can be rationally modulated during bioprocess production. PHAs have very similar physical properties to conventional polymers and can potentially replace them in various applications. They are thermoplastic and easily processable, their industrial application does not require significant changes in the usual equipment used for processing and molding of thermoplastics from petroleum, and they can be recycled or incinerated without toxic product generation. The CO_2 generated during the burning process is completely renewable. These facts, together with the growing demand for environmentally-friendly materials, have contributed to a reduction in cost and the expansion of the bioplastics trade.

PHAs' physical properties and their applications depend largely on their monomer composition and chain size, which depend on the substrates provided and the microorganism employed. Therefore, it is possible to set the biopolymers' properties obtained from operational process data, such as type and flow of substrates, fermentation time and microorganism strains.[9]

9.2.1.1   *Fundamental Bioprocess Steps for Obtaining PHAs*

 - Bioreactor and accessories' sterilization;
 - Growth medium sterilization;
 - Inoculum growth and adaptation in a small bioreactor;
 - Scale up to main bioreactor;

- Growth biomass process under aerobic conditions without nutrients and carbon source limitation;
- Biopolymer (PHA) production process can be anaerobic or limited-oxygen aerobic (low $_{KLa}$), with limited essential nutrients (one or more, such as Mg, N, P, S *etc.*) and an excess of the carbon sources (carbohydrates and carboxylic acids);
- Separation of biomass containing intracellular PHA;
- Cell disruption (solvents, hydrocyclone, ball mill *etc.*);
- Biopolymer separation, purification, concentration and drying;

Many gram-positive and gram-negative microorganisms, found in the soil, sea and effluent, are able to accumulate PHAs. The production costs of these plastics are directly related to microorganism type and substrates employed. It is desirable that producing strains have a high specific growth rate and are able to utilize low cost substrates with a good yield and are able to utilize effluents from food industries. The conversion factor substrate–PHA should be high and the ratio (PHA accumulated mass) –(total dry biomass) should be as high as possible.

Typically, P(3HB) production on an industrial scale uses gram-negative bacteria, such as *Cupriavidus necator, Alcaligenes latus* and recombinant *Escherichia coli*, because these microorganisms present a very good yield and easy growth using low cost substrates. However, P(3HB) isolated from gram-negative bacteria incurs additional costs during the purification steps, once the biopolymer granules are involved in lipopolysaccharide membranes, they may contain endotoxins, resulting in severe immunological reactions in the body, which would prevent this material's use for biomedical reasons. The additional costs relating to the separation and purification steps of P(3HB) can be avoided by working with gram-positive microorganisms.

The selection of mixed cultures with a high capacity for PHA accumulation occurs naturally because of reactor operation conditions and, consequently, there is no need to sterilize the system. On the other hand, having mixed cultures facilitates the use of complex substrates made from organic residues, such as food industry effluent, and the microbial population continuously adapts to substrate changes. Therefore, it is possible to minimize costs by using mixed cultures and selected substrates. The cost of PHAs produced by mixed cultures may decrease to around half of the cost of those produced by pure cultures, due to lower substrate and investment costs.

The following bacteria: recombinant *Ralstonia eutropha, Ralstonia eutropha*, recombinant *Escherichia coli, Burkholderia sacchari, Burkholderia cepacia, Azotobacter vinelanddi, Pseudomonas olevorans, Methylobacterium organophilum* and *Bacillus cereus* have more favorable characteristics for industrial scale production.

Among these microorganisms, *Cupriavidus necator* (*Ralstonia eutropha*), *Azotobacter vinelandii* and recombinant *Escherichia coli* are the most popular due to their yield and high biopolymer formation rate. *Ralstonia eutropha* is now called *Cupriavidus necator*. Most biodegradable polymers show hydrolyzable groups such as amide, ester, rhea, retane and others. *Pseudomonas* may produce PHAs from substrates containing alkanes and/or alkanoic acids. These PHAs may possess medium and long chains.[9]

The most popular PHA is P(3HB) and it can be synthesized by various microorganisms. The production of P(3HB) by *Burkholderia sacchari* and *Burkholderia cepacia* using hydrolyzate sugarcane bagasse has been accomplished with good yields. These strains were found in Brazilian sugarcane grounds in the state of São Paulo and their performance is comparable to *Alcaligenes eutrophus*, which is used as a reference for PHA production.

During P(3HB) production, sugar is hydrolyzed by an enzymatic process from *Alcaligenes sp.* bacteria. These bacteria are the same ones that produce PHB when subjected to nutritional stress conditions. There are other bacteria belonging to this genus that are also capable of hydrolyzing sucrose. Sugar can also be hydrolyzed through an enzymatic process in a reactor. This hydrolyzed sugar can be administered to producing microorganisms. During the process, a higher alcohol is used as a solvent for biopolymer extraction.

Sugarcane bagasse is used to produce electricity and steam. However, it can also be used to obtain products with higher added value *via* hydrolysis/physical–chemical treatment. The effluents are basically water and organic material, released over sugarcane tillage as organic fertilizer.

A good representative microorganism from the group of PH3B producers is *Ralstonia eutropha*. During balanced growth conditions, this bacterium catabolizes carbohydrates to pyruvate, *via* the Entner–Doudoroff pathway. Pyruvate can be converted by decarboxylation to acetyl-CoA. During reproductive growth, acetyl-CoA enters the citric acid cycle, releasing coenzyme A (CoASH) and is oxidized to CO_2 and water at the end, generating energy (ATP), reducing equivalents (NADH, NADPH and $FADH_2$) and biosynthetic precursors (oxaloacetate and 2-oxoglutarate). At this stage, there is no nutrient shortage and oxygen is dissolved at an ideal concentration for metabolism and cell reproduction. Therefore, the microorganisms do not need to accumulate biopolymer. It is a precursor step for biomass growth.

The biopolymer poly-3-hydroxybutyrate (P3HB) and its copolymer with 3-hydroxyvalerate (P3HB-*co*-3HV) are of interest because they are biodegradable and biocompatible. In Brazil, they are produced by bacteria from sucrose and propionate. Pereira conducted a study on cloning genes of 2-methylcitrate, which is linked to propionate catabolism synthesis in *Burkholderia sacchari* (a P3HB-*co*-3HV producer), and managed to obtain mutants by homologous recombination, which is more efficient for P3HB-*co*-3HV production.[10]

E. coli and *Clostridium* obtain gluconic acids from intermediary metabolism by this pathway from glucose.[9]

(1) From each glucose molecule, this pathway produces two NADPH molecules and one ATP molecule to be used in cellular biosynthetic reactions.

(2) Bacteria that possess this pathway enzyme can metabolize glucose without glycolysis or the pentoses pathway.

(3) This pathway is found in many bacteria including *Ralstonia*, *Rhizobium*, *Pseudomonas* and *Agrobacterium*.

Direct oxaloacetate amination and transamination lead to the synthesis of amino acids, which are incorporated into polypeptide cellular protein chains. The acetyl-CoA admission rate into the tricarboxylic acid cycle is dependent on the availability of nitrogen sources, phosphorus and other nutrients, dissolved oxygen concentration and the corresponding environment oxidative potential. Therefore, cell growth needs nutrients and oxygen, with an optimal concentration to metabolise and reproduce. If the cell is in an environment of limited essential nutrients for metabolism, it starts the production of energy and carbon reserves and the reproductive process is disrupted. From this moment, it begins the polymer production phase through the P(3HB) synthetic pathway.

Physical and mechanical methods are more suitable for cell disruption in general, because these methods cost less and do not affect intracellular biopolymer chemical integrity.[9] Depending on the chemical method, temperature and material, it may affect chemical integrity if no judicious choice of solvents and operating conditions is made. Physical methods can be ultrasound, hydrocyclone, mill balls, Hughes press or osmotic pressure.

9.2.1.2 Degradation

PHAs are degraded in aerobic and anaerobic systems, however, it is recommended that their degradation always takes place in an aerobic environment. Biodegradation occurs by the action of natural extracellular enzymes secreted by common microorganisms. Under anaerobic conditions, biodegradation occurs faster and slower in a marine environment. In an aerobic environment, the final products are CO_2, water and humus. Under anaerobic conditions, biodegradation generates methane, which can be used as a fuel. However, this is not currently done because it would be economically unfeasible. Anaerobic degradation is also avoided due to the fact that methane is a gas that produces and contributes to the greenhouse effect.

9.2.2 Polyhydroxybutyrate (PHB)

PHB is a homopolymer comprising monomer units of four carbon atoms. The impetus for its study in Brazil occurred in the early 90s, at the launch of the project "Production of biodegradable plastics from sugarcane through biotechnological processes" that was developed in cooperation with the Biotechnology Laboratory at the São Paulo State Institute of Technological Research (IPT), Copersucar Technological Center and the Biomedical

Sciences Institute from the University of São Paulo (USP) associated with the Federal University of Paraíba (UFPB), supported under PADCT-Finep.[6]

This pioneering work, and subsequent work developed by other research groups, focused on the following key topics for low cost PHA production: (a) finding efficient microorganisms able to accumulate PHA; (b) PHA production using domestic low cost raw materials and renewable agriculture; (c) increasing productivity through microbial cultivation with high cell density; (d) search for alternative reactors; (e) product characteristics by molar mass modulating.

PHB's arrival on the market has been hindered by the very feature that makes it so desired: biodegradability. The problem is that its degradation is unpredictable. Moreover, it is too brittle. Scientists at Cornell University in the United States found that adding clay nanoparticles can solve both problems.

PHB is much more malleable and its biodegradability can be adjusted simply by controlling the amount of nanoparticles added. Compared with the original PHB, the new plastic has a higher tensile rate and biodegradability than the original compound. The "nanoclays" also allow this plastic to deteriorate faster, around seven weeks only. This degradation is now more predictable and adjustable, which is a great advantage for this plastic. The degradation depends on the amount of clay nanoparticles added.[9]

9.2.3 Polyhydroxyalkanoates with Medium Chain Monomers (PHAmcl)

From the advances seen in PHA production with short chain monomers (PHAscl), such as PHB, PHB-*co*-HV and PHB-*b*-HPE, it was evident how important it was to diversify PHA monomer composition, as these materials' properties are largely dependent on this characteristic.[6] Once monomer composition control is achieved using different tools, it is expected that PHAs can be tailored for different applications. For example, PHAmcl, a PHA composed of medium chain monomers (monomers with 6 to 14 carbon atoms) has different properties to PHAscl. While the short chain monomers resemble plastics, such as polyethylene, PHAmcl monomers present elastomeric properties, resembling rubber. They can be produced either from carbohydrates and various aliphatic compounds by microorganisms, especially bacteria belonging to the *Pseudomonas* genus, belonging to group I of ribosomal RNA homology.

Initially, it was expected that the use of fatty acids was the most interesting way to produce and control these materials' composition, once the polyester composition reflects the carbon chain length from the offered source. When the main source is nonanoic acid, for example, *Pseudomonas oleovorans* produces PHAmcl composed of 3-hydroxynonanoate predominantly. However, it was discovered in the 90s that a group of *Pseudomonas* accumulates PHAmcl from carbohydrates. When produced from carbohydrates, PHAmcl presents 3-hydroxydecanoic acid (HD) as the main constituent and other monomers can also be detected. In addition, comparisons of the minimum cost of carbon source for the production of PHAmcl from carbohydrates and soybean oil showed similar values.

Although the first studies of PHAmcl production date back to 1998, various bacteria had previously been isolated from Brazilian soil but it was only found later that the polymer produced by them was PHAmcl.

More recently, bacteria isolated from Atlantic Forest soil in Brazil were evaluated for the production of rhamnolipids and PHAmcl. Among the isolates, 26 have produced anionic biosurfactant and 57 have produced PHA. Some strains synthesized both compounds simultaneously.

Trials for finding PHAmcl producing strains from vegetable oils were also conducted with soy oil, rice, canola, corn and sunflower as carbon sources in a rotating incubator, revealing the presence of monomers containing unsaturations. Previously isolated bacteria capable of accumulating PHAmcl from carbohydrates were tested and compared with the reference strain *P. putida* KT2440.[6]

9.2.4 Polyhydroxy-pentenoate (PHPE)

Poly-3-hydroxy-4-pentenoate (PHPE) is a polymer consisting of short monomer units (five carbons) with an unsaturation in its side chain. It was the first unsaturated chain homopolymer described in the literature. Its physical properties are different from those presented by the homopolymer PHB and the copolymer PHB-*co*-HV. PHPE is less crystalline and has a lower melting temperature.

The isolation of *Burkholderia* sp TPI 064 (recently *B. cepacia*) was recently discovered. This bacterium is capable of accumulating polymers with unsaturated monomers from carbohydrates, *i.e.* without substrate supply with chemical structure directly related to the product. This strain, as well as ten other strains isolated from soil and poly(HB-*co*-HPE) producers, were compared with conversion factors from substrates to polymer (Y^{G}P/S). The best strains were evaluated with respect to maximum specific growth rate values (μ_{Xmax}) and for pentenoic acid conversion into HPE units.[6]

9.2.4.1 Biodegradation Evaluation

The various studied modifications and formulations involving PHAs require constant proof of the maintenance of one of their essential characteristics: biodegradability. One way of assessing biodegradability is to measure the PHA carbon conversion in carbon dioxide by some bacteria in the environment.

The ability of various microorganisms to degrade PHAs and other polyesters may involve one or more enzymatic mechanisms. Studies to isolate and to characterize PHA, poly-caprolactone (PCL) and polylactates (PLA) degrading microorganisms have showed microbial groups capable of degrading different materials. It may be relevant for rapid preliminary tests to evaluate the biodegradability of new developing formulations.

There are some reports stating that PHB and PCL when molded by hot compression become more susceptible to fungi attack. PHB's and PCL's biodegradability under composting at different pHs was also evaluated after irradiation with UV-B. Only one specific irradiation and pH condition increased biodegradation. For all cases, PHB showed a greater mass loss than PCL.[6]

It is essential to have a safe process during the whole bioplastic life cycle. PHB has been described as "the first real thermoplastic example obtained from a biotechnological process" that is actually biodegradable. Although PHB is stable in a natural environment, its degradability rate is very high under processing and fusion conditions.

Biopolymer production plant integration (PHB as an example) with a sugar mill offers unique advantages, from an economic perspective and for being an environmentally safe process, once sugar mills have energy availability (thermal and electric) from renewable energy sources—they are almost free in Brazil. For example, processing sugarcane bagasse uses an efficient management system, uses agricultural and industrial waste, uses extensive knowledge and the availability of large-scale fermentation processes. It also uses natural and biodegradable solvents.[5]

9.2.5   PHA Use in Food Packaging

The main functions of packaging are to contain, to protect, to be convenient, to communicate and to sell a product.[11,12] These functions' definitions are as follows:

(i)   Containment: this is the basic function of the package, to contain a certain amount of food, unitizing the product and facilitating its transportation, storage, sale and use.

(ii)   Protection: this is the primary function of the package, acting as a barrier between the food and the external environment, protecting it against environmental factors that could accelerate product degradation.

(iii) Convenience: modern industrialized societies have brought about enormous changes in life styles and the packaging industry has had to respond to those changes. Today, packaging is an increasingly versatile and functional tool for users and consumers.

(iv)   Communication: the package may have useful information for consumers, such as product identification, amount and composition, directions for use, warnings (allergens) and sustainability information.

(v)   Selling: the first visual contact with the product is through the packaging and it has an important effect on the consumers' purchase decision. In addition, the packaging should be attractive and allow rapid product identification by the consumers.

According to these authors, the most important function is protection once it has directly addressed consumer safety.[11,12] Within the protection function context, barrier properties against environmental actions such as light, humidity and O_2 represent a key role in food stability during the storage period. The packaging barrier properties are critical for the preservation of many foods. Therefore, appropriate packaging contributes to the delay of undesirable changes, preventing the migration of gases, moisture, fat and other volatile materials, and provides support against mechanical stress.[13]

The suitability of product packaging minimizes undesirable changes, increasing food stability. However, depending on the time of product–packaging contact, there will be interactions (except for glass containers, which do not interact with food). The packaging's compatibility with food reduces interactions, but does not avoid them totally.

Other requirements for packaging materials include processing capacity, ability to be heat-sealed, ability to be recycled or composted, and low cost.[13]

According to Anjos,[14] the most prominent plastics in the food industry are: (i) low density polyethylene (LDPE), which is inexpensive and is used in dehydrated products in general, pasteurized milk and laminated packaging; (ii) high density polyethylene (HDPE)—its largest use is in bags for supermarkets; (iii) polystyrene (PS), used in the preparation of expanded trays for various food products; (iv) poly(vinyl chloride) (PVC), used for packaging vinegar, mineral water, edible oils, sauces and wrappers for food *etc.*; (v) polypropylene (PP), used for high fat food, it can be oriented and bi-oriented and shows excellent gloss and transparency; (vi) poly(ethylene terephthalate) (PET) is mainly used in carbonated beverages and (vii) poly(ethylene naphthalate) (PEN), which has more specific applications than PET due to its high price. PEN is found in fruit juices, preserves, beers, wine and returnable packaging.

The plastics industry has been growing steadily over the last 60 years. World production increased from 1.7 million tons in 1950 to 265 million tons in 2010. Worth noting is the exponential increase between the 1950s and the 1970s, and it is possible to observe how polymer production doubled from the 1970s to the 1990s. This production value almost tripled in the 2000s. Despite the global economic crisis and the fall in the consumption and production of plastics in 2008 and 2009, the market has recovered and plastics production was at a record high in 2010, with a 6% increase in production from 2009. Among the world's producers, the largest was China in 2010, with 62 million tons approximately, followed by the EU, which produced 57 million tons approximately.[15]

Referring to the latter producer, the demand for processed plastics was 46.4 million tons in 2010. Total plastic production demands were higher than in 2009 (55 and 45 million tons, respectively), but they have not yet reached the 2007 numbers (pre-global economic crisis), when the EU produced 60 million tons of plastics and demand was 48.5 million tons.

Regarding consumption by segment, plastic packaging leads with a demand for 18 million tons, followed by construction (9.55 million tons), motor vehicles (3.5 million tons) and electronic equipment (2.6 million tons).[15]

The main raw material for plastics manufacture is oil, but only 4% of the world's oil and gas production is used for plastics production and 3–4% is used as energy in the process.[16,17] The plastics industry is constantly developing, dealing with new technologies to answer new demands, and it is no surprise that the world production of plastic reached 265 million tons in 2010.[15]

Plastic polymers can be classified into two distinct groups by thermal behavior during processing: thermoplastic and thermosetting. Thermoplastics are moldable, as they soften when heated. This process can be repeated numerous times and polymer degradation is minimal. On the other hand, thermosets are not easily moldable by heating. During processing, these polymers are moldable but become rigid when the process is done and they become resistant to temperature increase. Thermosetting plastics are typically stiffer than thermoplastics.[18]

The main thermoplastic polymer types are acrylic, cellulosic, ethylene vinyl acetate (EVA), polyethylene terephthalate (PET), polyamides (nylons), polyethylene (PE), polystyrene (PS), polyvinyl chloride (PVC), polycarbonate and polypropylene (PP).[18] PET, PVC, HDPE, LDPE, PS and PP have a higher production volume and relatively low cost.[19] The main thermosetting polymer types are aminoplastics, epoxies, phenolics (phenol formaldehyde), polyesters and silicones.[18] These polymers have a wide range of applications and their features are very different.

Despite the many benefits that plastic brings to human society, their residues are harmful. The large plastic volume, the huge amount of post-consumer disposal and environmental impact caused by the improper disposal of waste that is not biodegradable, are just some of the problems. In addition, plastics can damage human and animal health, mainly because of the additives and chemicals used in their manufacture. Regulatory instruments to mitigate plastics' effects on human health and the environment need to follow their cycle from production and use to disposal since most of the plastic items sold (especially packaging and non-durable goods) become residues in less than a year or after a single use (worst scenario). However, plastic wastes are valuable raw material sources, which can be converted into energy or other polymeric materials.[20] One of the challenges is to increase recycling programs. About 31.04 million tons of plastic waste were produced in the United States in 2010 and only 2.64 million tons were recycled (7.6%).

There is good evidence that by increasing recycling potential, there will be significant growth in the amount of post-consumer products that can be recycled.[21,22] Currently, fossil fuel consumption for plastics production is linear, from oil to disposal, but it must become cyclical, with further recovery to become a raw material again.[17]

The high cost and scarcity of landfill space for the creation of new ones have influenced the development of alternative techniques for waste disposal. The impact of plastics is even more serious when plastics are disposed of improperly and dispersed in the environment. This has led to techniques for recycling and incineration and these techniques have been more usually practiced.

According to Christiana Figueres, the Executive Secretary of the United Nations Framework Convention on Climate Change (UNFCCC), in a statement issued in Bonn, Germany, where the UNFCCC was held in May 2013, the concentration of carbon dioxide (CO_2) in the atmosphere surpassed 400 parts per million (ppm) for the first time, leaving the planet in a "danger zone". Statements like these, ineffective policies for the treatment of solid waste and a very low recycling rate, make the population more concerned about the environment. In this context, the use of materials with environmental appeal has been well regarded.

The properties required for plastic packaging depend mainly on the changes shown by the product and the conditions where it is stored. For food packaging applications, bioplastics should: (i) have good sensory properties, a gas and mechanical barrier; (ii) have sufficient biochemical, physical–chemical and microbiological stability; (iii) be free of toxic waste and be safe for consumption; (iv) have simple manufacturing technology; (v) not pollute; (vi) have good availability and low cost (raw material and process).[23]

The main natural biomaterials for bioplastics are proteins, cellulose derivatives, alginates, pectins, starch and other polysaccharides. The solubility in water of the polysaccharide film is advantageous in situations where the film is consumed with the product, resulting in little change in the food's sensory properties.[24] Edible films based on proteins, polysaccharides or lipids minimize special care with the final package and increase food quality.[25]

9.2.5.1 Methods for Obtaining Plastic Products

Various methods are used in plastic product manufacture, but the four main ones are extrusion, injection, blow and rotational.[26]

According to Smith,[27] the extrusion process has many advantages such as versatility, high quality of product, low production cost, the formation of products with different geometric shapes and no formation or release of effluents. This process can be divided into three stages: pre-extrusion, extrusion and post-extrusion, where the equipment for the first and last steps depends on the type of material to be used, while in the extrusion step, processing conditions may vary.[28]

The greatest technique for converting thermoplastics is injection molding and is widely used for bottle and jar production, as well as production of caps and dispensers.[12] Various plastic accessories are also manufactured using this technique.

The injection molding process consists essentially of three consecutive steps, starting with supply and softening of the thermoplastic material in a heated cylinder and its subsequent injection, under high pressure, within a relatively cold mold, where the material gets hard and takes the final shape. Finally, the molded article is cooled and expelled from the mold by ejector pins, compressed air or other auxiliary equipment.[12]

The resins that are commonly injection molded include high and low density polyethylenes, polypropylene and polyethylene terephthalate; the latter as beverage bottle preforms that are subsequently blown.

Compared to extrusion, injection molding is presented as a cyclical process. A complete cycle consists of the following operations, starting with plastic granulate material dosage in the injection cylinder, followed by fusion of this material until it gets injection consistency. After the fusion, the molten plastic material is injected into the closed mold and this injected material is cooled until its solidification. Finally, the product is extracted with the opened mold.[12]

The potential use of injection molding with biopolymers is great despite it not being used much yet. Félix *et al.*[29] evaluated injection molding as an alternative to biomaterial production from prepared mixtures using different proportions of albumin/soybean and glycerol as a plasticizer. Viscoelastic measurements and differential scanning calorimetry (DSC) of different protein mixtures and glycerol were used to select the most suitable processing conditions. After evaluating the physico–chemical properties, the authors concluded that both proteins and their mixtures produced bioplastics by injection molding, although with lower mechanical properties than standards made of low density polyethylene (LDPE).

Usually, compression molding is studied as a precursor of extrusion,[30] to study the material machinability and to identify appropriate conditions for extrusion. In this study, the authors have shown that the mechanical properties of whey protein sheets molded by compression with 40–50% plasticizer were better than films produced by casting with 45% plasticizer. The authors also claim that extrusion is a quick process, requires little space and has a small number of production steps.[31]

9.2.5.2 Examples of the Application of Bioplastics in Food Packaging

The applications of biopolymers in articles for contact with food include disposable cutlery, cups, salad pots, plates, films for wrapping and lamination, straws, stirrers, lids, plates and food containers for delicatessen and fast food restaurants. These articles have contact with aqueous solutions, acid food and fatty foods. They are kept at room temperature or below, also at high temperatures (60 °C), followed by cooling to room temperature or below.[32]

PHB can be used in injection molding processes for the manufacture of food packaging, with the same equipment used for PP packaging injection. However, the process conditions should be adjusted according to polymer characteristics.[33] The authors found a notable difference between PHB and PP bottles in relation to their performance in dynamic compression resistance and a drop test, in that PHB was as hard as PP, but less flexible. They also concluded that PHB performance tended to be worse than PP performance at refrigerator and freezer temperatures. However, PHB performance was better at higher temperatures. The physical, dimensional, mechanical and sensory tests showed that PHB can replace PP containers for food products with high fat content (mayonnaise, margarine and cream cheese), including storage in freezers and heating in microwave ovens Muizniece-brasava *et al.*[34] reported that PHB materials are suitable for sour cream storage.

Basic PHB has relatively high glass transition and melting temperatures. To improve flexibility for potential packaging applications, PHB is synthesized with various co-polymers such as poly (3-hydroxyvalerate) (HV), leading to a decrease in the glass transition and melting temperatures. In addition, the HV broadens the processing window since there is improved melt stability at lower processing temperatures.[35]

Fabra *et al.*[36] have created an innovative way to develop renewable bio-polyester microbial-based multilayer structures with enhanced barrier performance of significant interest for food packaging applications. These researchers developed multilayer structures based on polyhydroxybutyrate-*co*-valerate with a valerate content of 12% (PHBV12) containing a high barrier interlayer of zein electrospun nanofibers. It was observed that the method used for the preparation of the outer layers affected PHBV12's functional properties, since increased mechanical resistance and water vapor permeability and transparency decreased in the multilayer containing outer layers prepared by compression molding to a higher extent than for their counterparts obtained by casting. The addition of a zein minimum interlayer produced changes in the mechanical and optical film properties while the incorporation of the zein nanostructured interlayers significantly improved the oxygen barrier properties of the multilayer films prepared by both processing technologies. However, the effect of the interlayer on the water vapor permeability and limonene depended on zein content.

The incorporation of 3-hydroxyvalerate (HV) in PHB, resulting in PHBV, increased impact strength, elongation modulus, tensile strength and decreased the Young's modulus, making the film more flexible and more resistant.[35] The price is very high but PHBV degrades between five and six weeks in a microbiologically active environment, resulting in water and carbon dioxide under aerobic conditions. In an anaerobic environment, degradation is faster, producing methane.[32]

A mixture of PHBV with PLA had a positive effect on the elasticity modulus, elongation at break and flexural strength for different blends. However, tensile strength did not improve in any of them. In the same way, Zhang *et al.*[37] reported improved mechanical properties for blends of PHB/PLA compared with the common PHB. In addition, PVA (polyvinylacetate) grafted on PIP (poly-*cis*-1,4-isoprene) and mixed with PHB had better tensile properties and impact strength than PHB/PIP blends, which were immiscible.[35] Combined with synthetic plastics or starch, PHAs make excellent packaging films.[38]

Levkane *et al.*[39] have investigated the pasteurisation effect in a meat salad with conventional packaging (PE, PP) and packaging with a biopolymer (PLA, PHB) and have found that PHB films can be used successfully to pack this type of food. Haugaard *et al.*[40] have found that orange juice and sauce packaged in PHB resulted in the same quality changes when compared to the same products packaged in HDPE. This study has shown that juices, other commercial acid beverages, sauces and other fatty foods can be packed in PHB.

In another study, conducted by Shen *et al.*,[41] thermoplastic starch (TPS) blended with PHA had a positive effect on the barrier and hydrolytic properties and UV stability of a starch-based film. With this blend, it was possible to reduce the processing temperature, resulting in less degradation of the starch.[41]

Biopolymers satisfy environmental concerns, but they may have some limitations in terms of performance, such as heat resistance, barrier properties and mechanical properties associated with costs. Some PHA/PHB films may have increased fragility (due to high glass transition temperatures and melting temperatures), greater stiffness, less impact resistance and less heat resistance. All these factors are limiting the application of these films in food packaging.[42–44]

9.3 Conclusion

Over the last few decades, polymers have replaced conventional materials (metals, ceramics, paper) in packaging applications. The use of synthetic polymers in food packaging is growing more and more, where they provide mechanical, chemical, and microbial protection from the environment and allow product display. Now, concerns about the environment and the need to reduce the dependency on fossil fuels has led to the search for alternatives to traditional packaging materials, such as biodegradable materials (PHB, PHB-*co*-HV, PHB-*b*-HPE, and others). Therefore, these types of packaging materials need more research to result in products with higher added value, such as research into the introduction of smart molecules able to give information about the properties of the food inside the package and its nutritional value. It is necessary to research these types of materials to improve barrier properties, to ensure integrity of foods, to incorporate labels and intelligent packaging, giving the consumer the opportunity to have more detailed information about the product compared to current packaging systems.

References

1. A. B. Reis, *Caracterização de filmes e coberturas de quitosana aplicados em papelão ondulado*, Dissertação de Mestrado, Unicamp-FEQ, Campinas-SP, 2005, p. 74.

2. ASTM D6866-06, *Standard Test Methods for Determining the Biobased Content of Natural Range Materials Using Radiocarbon and Isotope Ratio Mass Spectrometry Analysis*, American Society for Testing and Materials, 2006, p. 14.
3. ASTM D6400-04, *Standard Specification for Labeling of Plastics Designed to be Aerobically Composted in Municipal or Industrial Facilities*, American Society for Testing and Materials, 2004, p. 3.
4. A. U. B. Queiroz and F. P. Collares-queiroz, *Polym. Rev.*, 2009, **49**, 65.
5. L. Velho and P. Velho, *Brasília*, 2008, **6**, 225.
6. L. F. da Silva, J. G. C. Gomez, R. C. S. Rocha, M. K. Taciro and J. G. da C. Pradella, *Quím. Nova*, 2007, 7, 30.
7. M. A. de Moraes, Obtenção e caracterização de blendas e compósitos poliméricos de fibroína de seda e alginato de sódio, Dissertação de Mestrado, Unicamp-FEQ, Campinas-SP, 2010, p. 101.
8. C. Marangoni, A. Furigo Jr and G. M. F. Aragão, *Biotechnol. Lett.*, 2000, **22**, 1635.
9. A. A. Amates, G. V. Barros and V. F. Yamanishi, *Biopolymers*, 2009, **16**, 10.
10. E. M. Pereira, *Clonagem de genes do ciclo de 2-metilcitrato e avaliação de estratégias para a obtenção de mutantes prp de "Burkholderia sacchari" produtora de plástico biodegradável*, Dissertação de Mestrado, IPT, USP, São Paulo-SP, 2003, p. 99.
11. K. L. Yam, Emerging Food Packaging Technologies: an overview *in Emerging Food Packaging Technologies: Principles and Practice*, ed. K. L. Yam and D. S. Lee, WoodHead Publishing, Cambridge, 2012, p. 512.
12. G. L. Robertson, *Food Packaging, Principles and Practice*, CRC Press, Boca Raton, 2nd edn, 2006, p. 550.
13. K. S. Mikkonen and M. Tenkanen, *Trends Food Sci. Technol.*, 2012, **28**, 90.
14. C. A. R. Anjos, *Importância da embalagem na conservação e distribuição de alimentos*, Apostila da disciplina Tecnologia de Embalagens, do programa de pós graduação da FEA-UNICAMP, Campinas, SP, 2004, p. 30.
15. *Plastics – the Facts 2011*. An analysis of European plastics production, demand and recovery for 2011, Plastics Europe.
16. J. Hopewell, R. Dvorak and E. Kosior, *Philos. Trans. R. Soc., B*, 2009, **364**, 2115.
17. R. C. Thompson, C. J. Moore, F. S. Saal and S. H. Swan, *Philos. Trans. R. Soc., B*, 2009, **364**, 2153.
18. R. A. Parente, *Elementos estruturais de plástico reciclado*, Dissertação de Mestrado, USP, São Paulo-SP, 2006, 142.
19. A. L. Andrady and M. A. Neal, *Philos. Trans. R. Soc., A*, 2009, **364**, 1977.
20. S. M. Al-Salem, P. Lettieri and J. Baeyens, *Prog. Energy Combust. Sci.*, 2010, **36**, 103.
21. J. C. Michaud, L. Farrant and O. Jan, *Wrap-Waste and Resources Action Programe: Environmental Benefits of Recycling, Bio Intelligence Service and Copenhagen Resource Institute-Final Report*, Oxon, Copenhagen, 2010, p. 255.
22. S. M. Al-Salem, P. Lettieri and J. Baeyens, *Waste Manage.*, 2009, **29**, 2625.

23. F. Debeaufort, J.-A. Quezada-Gallo and A. Voilley, *Crit. Rev. Food Sci.*, 1998, **38**, 299.
24. I. G. Donhowe and O. Fennema, *Edible coating and films to improve food quality*, ed. J. M. Krochta, E. A. Baldwin, M. Nisperos-Carriedo, R. Hagenmaier and J. Bai, CRC Press, Technomic, Lancaster, 1994, p. 392.
25. H. Chen, *J. Dairy Sci.*, 1995, **78**, 2563.
26. American Chemistry Council, 2011. Available in http://www.americanchemistry.com/Safety/ProductSafety. Acessed in 29/12/2013.
27. O. B. Smith, *Cereal Foods World*, 1976, **21**, 4.
28. A. El-Dash, *A renewable resource: theory and practice*, ed. Y. Polymeronz and L. Munch, AACC International, St Paul, Minnesota, 1982, p. 728.
29. M. Félix, J. E. Martín-alfonso, A. Romero and A. Guerrero, *J. Food Eng.*, 2014, **125**, 7.
30. R. Sothornvit, C. W. Olsen, T. H. McHugh and J. M. Krochta, *J. Food Eng.*, 2007, **78**, 855.
31. R. Sothornvit, C. W. Olsen, T. H. McHugh and J. M. Krochta, *J. Food Eng.*, 2007, **78**, 855.
32. V. Siracusa, P. Rocculi, S. Romani and M. Dalla Rosa, *Trends Food Sci. Technol.*, 2008, **19**, 634.
33. D. Z. Bucci, L. B. B. Tavares and I. Sell, *Polym. Test.*, 2005, **24**, 564.
34. S. Muizniece-brasava and L. Dukalska, *Proceedings of the Latvia University of Agriculture*, 2006, **16**, 79.
35. S. J. Modi, Thesis: *Assessing the Feasibility of Poly-(3-Hydroxybutyrate-co-3-Valerate) (PHBV) and Poly-(Lactic Acid) for Potential Food Packaging Applications*, PhD Thesis, The Ohio State University, Ohio, 2010, p. 105.
36. M. J. Fabra, A. Lopez-rubio and J. M. Lagaron, *Food Hydrocolloids*, 2013, **32**, 106.
37. L. Zhang, C. Xiong and X. Deng, *Polymer*, 1996, **37**, 235.
38. R. N. Tharanathan, *Trends Food Sci. Technol.*, 2003, **4**, 71.
39. L. Dukalska, E. Ungure, I. Augspole, S. Muizniece-Brasava, V. Levkane, R. Tatjana and I. Krasnova, *Proceedings of the Latvia University of Agriculture*, 2014, **30**, 22.
40. V. K. Haugaard, B. Danielsen and G. Bertelsen, *Eur. Food Res. Technol.*, 2003, **216**, 233.
41. Z. Shen, G. P. Simon and Y. B. Cheng, *Polymer*, 2002, **43**, 4251.
42. N. Peelman, P. Ragaert, B. De Meulenaer, D. Adons, R. Peeters, L. Cardon, F. Van Impe and F. Devlieghere, *Trends Food Sci. Technol.*, 2013, **32**, 128.
43. L. Shen, J. Haufe and M. K. Patel, Product overview and market projection of emerging bio-based plastics, 2009. http://en.european-bioplastics.org/wpcontent/uploads/2011/03/publications/PROBIP2009_Final_June_2009.pdf Accessed 12.04.2012.
44. J. S. Yoon, W. S. Lee, H. J. Jin, I. J. Chin, M. N. Kim and J. H. Go, *Eur. Polym. J.*, 1999, **35**, 781.

Subject Index